Python 在语言研究中的应用

张鹏华　潘一　编著

西安电子科技大学出版社

内 容 简 介

本书从 Python 的安装与管理入手,从 Python 初学者的角度为读者呈现 Python 基础及其在语言学习、研究和翻译实践中的应用,以学习、研究和机器翻译实践中的具体问题为导向,给出解决这些问题的代码及注释,并将有关内容可视化,涉及语言学习、研究中的词汇、句法和语义方面以及国内外不同类型最新的机器翻译等内容。书中案例深入浅出,突出应用,旨在提高读者解决实际问题的能力。全书共 9 章,主要包括 Python 安装与管理、Python 编辑器、Python 基础、自然语言处理工具包、句法分析、正则表达式、文本处理、情感分析和机器翻译等,涵盖语言学习、研究的大部分领域。

本书既可为读者学习编程提供参考,也可为语言学习者、从业者提供一些研究方法和新的研究思路。

图书在版编目(CIP)数据

Python 在语言研究中的应用 / 张鹏华,潘一编著. -- 西安:西安电子科技大学出版社, 2025. 1. -- ISBN 978-7-5606-7504-6

Ⅰ. TP312.8;TP391

中国国家版本馆 CIP 数据核字第 20246R76P2 号

策　　划　　刘小莉
责任编辑　　刘小莉
出版发行　　西安电子科技大学出版社(西安市太白南路 2 号)
电　　话　　(029)88202421　88201467　　　邮　　编　　710071
网　　址　　www.xduph.com　　　　　　　　电子邮箱　　xdupfxb001@163.com
经　　销　　新华书店
印刷单位　　陕西天意印务有限责任公司
版　　次　　2025 年 1 月第 1 版　　2025 年 1 月第 1 次印刷
开　　本　　787 毫米×1092 毫米　　1/16　　印　张　16.5
字　　数　　389 千字
定　　价　　48.00 元

ISBN 978-7-5606-7504-6

XDUP 7805001-1

*** 如有印装问题可调换 ***

前　言

在语言学习、研究的过程中不可避免地要处理语言数据，语言学习和研究人员往往会使用现成的相关软件完成工作。但也会遇到无法用现有的软件与工具快速、便捷地完成任务，而不得不采用耗时、费力的其他方法的情况。另外，在语言学习、研究中涉及文本处理、文本搜索、比较、提取、统计、可视化等多项任务时，很难使用某一种软件完成多项任务。

实际上利用 Python 语言即可解决上述提及的大部分问题。Python 语言简洁明了，便于初学者学习和使用，可以完成多项语言处理、分析、研究任务，也可应用于语言学、翻译学习研究的多项任务中。

为便于语言学习者和从业人员学习，本书刻意弱化了理论知识的讲解，侧重应用程序展示，力求让读者尽快学以致用。本书从 Python 的安装、编辑器的使用入手，依次介绍了 Python 语言基础、自然语言处理工具包、特定信息提取与分析、句法分析、正则表达式、文本处理、情感分析及其可视化和机器翻译。同时，本书力争引入最新相关内容及方法，希望为学习者提供最新的研究方法和程序。

本书提供了常用的信息提取、情感分析和机器翻译的 Python 代码，同时调用了最新的 GPT 模型 API，以便高效提取关键词、分析情感并进行文本翻译。为了让读者更好地理解代码逻辑，代码中添加了详细的注释，并展示了代码运行结果以供参照。这些关键词提取、句法分析、情感分析以及机器翻译的方法，不仅是语言学习与研究的热点，也代表了当前语言研究领域的最新发展趋势。通过本书的指引，读者能够轻松掌握这些前沿技术，进而深化对语言学、翻译的理解与研究。

全书由西安电子科技大学张鹏华老师、西安交通大学潘一老师合作完成。其中张鹏华老师负责全书统稿与第 3 章至第 9 章的编写，以及全书的代码编写和测试等工作；潘一老师负责第 1、2 章的编写工作。

感谢西安电子科技大学计算机学院王书振教授等多位老师的帮助以及西安电子科技大学科学研究院和出版社的资助。作者在编写本书时得到了许多学者的支持和鼓励，同时参考和借鉴了有关专家和教研人员的研究成果，吸取了西安电子科技大学出版社编辑提出的建设性意见，在此一并对他们所付出的辛勤劳动表示诚挚的感谢。

由于作者水平有限，书中难免存在欠妥之处，敬请广大读者提出批评和建议，以便日后修订改进。

作　者

2024 年 8 月

目 录

第 1 章 Python 安装与管理 ... 1
1.1 Python 和 Anaconda 的安装 ... 1
1.1.1 使用 Python 程序包安装 ... 1
1.1.2 安装 Anaconda ... 3
1.2 环境变量配置 ... 14
1.3 Conda 运行环境及管理 ... 17

第 2 章 Python 编辑器 ... 23
2.1 常用编辑器概况 ... 23
2.2 Jupyter 安装与使用 ... 24
2.3 NotePad 安装与使用 ... 25
2.4 Sublime Text 的配置与使用 ... 27
2.5 PyCharm 的安装与使用 ... 32

第 3 章 Python 基础 ... 39
3.1 Python 常识 ... 39
3.2 变量和数据类型 ... 41
3.3 数值 ... 44
3.4 字符串 ... 47
3.5 列表 ... 53
3.6 字典 ... 59
3.7 集合 ... 66
3.8 元组 ... 70
3.9 推导式 ... 74
3.10 函数 ... 78

第 4 章 自然语言处理工具包 ... 88
4.1 NLTK 概况及安装 ... 88
4.2 NLTK 语料库及资源 ... 93

4.3	语料处理	111
4.4	文本分类	129
4.5	NLTK 自定义语料库	137
4.6	文本特征统计	139

第 5 章 句法分析 ... 145
5.1	NLTK 句法分析	145
5.2	Stanford CoreNLP 句法分析	153
5.3	可视化句法分析	159

第 6 章 正则表达式 ... 161
6.1	正则表达式的概念、用法及用途	161
6.2	Python 中使用正则表达式	162
6.3	使用正则表达式爬取网页数据	169
6.4	常用的正则表达式	171

第 7 章 文本处理 ... 174
7.1	提取中、英文	174
7.2	提取词汇和短语	174
7.3	提取句型	179
7.4	提取特定特征文本	181
7.5	提取关键词	185
7.6	提取文本摘要	205
7.7	词云	209

第 8 章 情感分析 ... 214
8.1	情感分析原理	214
8.2	TextBlob 情感分析	214
8.3	SnowNLP 文本情感分析	217
8.4	VADER 文本情感分析	221
8.5	GPT 情感分析	225

第 9 章 机器翻译 ... 227
9.1	Marian 机器翻译	227
9.2	GPT 机器翻译	228
9.3	百度翻译服务	232

9.4 Meta 机器翻译 238
9.5 Google 翻译 243
9.6 DeepL 翻译系统 250
9.7 讯飞星火认知模型翻译 252

参考文献 256

第 1 章 Python 安装与管理

1.1 Python 和 Anaconda 的安装

Python 程序包的安装通常有两种方法：一种是在 Python 官方网站下载安装程序进行安装；另一种是使用 Anaconda 安装 Python 程序包。

Python 官方网站下载界面如图 1-1 所示。

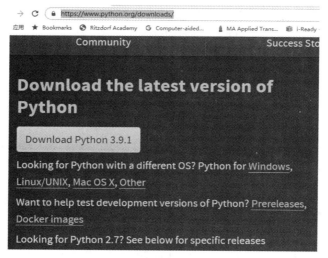

图 1-1 Python 官方网站下载界面

本书建议使用第二种安装方法，Anaconda 附带了一大批常用数据科学包，包括 Conda、Python 和多个科学包及其依赖项，便于安装、卸载和更新包以及配置、管理运行环境。

1.1.1 使用 Python 程序包安装

1. 下载 Python 安装包

以 3.6.5 版本为例，在 Python 的官网中找到 3.6.5 版本的 Python 安装包，点击进行下载。若系统是 32 位的，则选择 32 位的安装包；若系统是 64 位的，则选择 64 位的安装包。

2. 安装

(1) 双击下载好的安装包，弹出如图 1-2 所示的 Python 安装界面。要注意的是，安装时应将 Python 加入 Windows 的环境变量中。如果忘记勾选该选项，就需要手动加到环境变量中。这里选择的是自定义安装，点击"自定义安装"。

Python 在语言研究中的应用

图 1-2　Python 安装界面

(2) 进入 Python 安装组件选择界面后，选择需要安装的组件，点击"Next"，如图 1-3 所示。

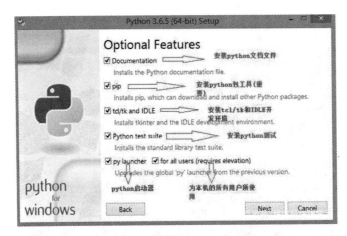

图 1-3　Python 安装组件选择界面

(3) 进入自定义安装路径界面，如图 1-4 所示。

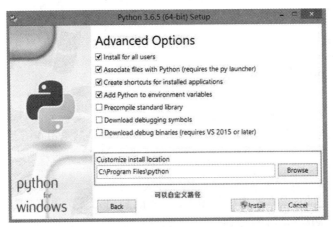

图 1-4　自定义安装路径界面

(4)点击"Install"后,系统开始安装,Python 安装进程界面如图 1-5 所示。

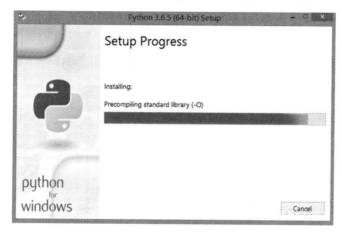

图 1-5　Python 安装进程界面

(5)安装完成后,出现安装成功的提示界面,如图 1-6 所示。

图 1-6　Python 安装成功界面

1.1.2　安装 Anaconda

Anaconda 可以便捷地获取并管理包,同时可以对环境统一管理。Anaconda 有包含 Conda、Python 等在内的 180 个包及其依赖项。Anaconda 分为个人版、商业版、团队版、企业版、专业版等几个版本,个人用户可选择安装个人版。根据自己的电脑操作系统(Windows、MacOX、Linux 系统)类型选择 32 位或 64 位版本安装。

Anaconda 是开源的,其安装过程简单,可以方便地使用 Python、R 语言,有免费的社区支持。上述功能的实现主要基于 Anaconda 拥有的 Conda 包、环境管理器和 1000 多个开源库。

日常工作或学习并不需要使用这么多库,此时可以考虑安装 Miniconda(这里不介绍 Miniconda 的安装及使用)。

Anaconda 可以在以下系统平台中安装和使用:Windows、MacOS 和 Linux(x86/Power8)。

Anaconda 的安装要求如下：

(1) 系统：32 位或 64 位系统均可。

(2) 下载文件大小：约 500 MB。

(3) 所需空间大小：3 GB(Miniconda 仅需 400 MB 空间即可)。

Anaconda 的下载界面如图 1-7 所示。

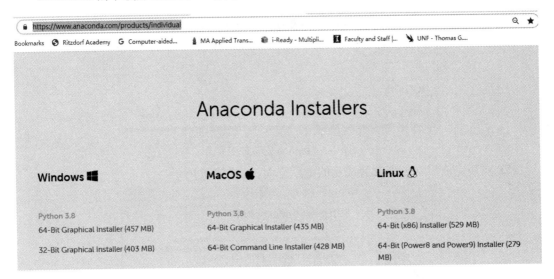

图 1-7　Anaconda 的下载界面

1. Windows 系统安装 Anaconda

(1) 根据前面所述的方法，按计算机硬件条件选择"64-Bit Graphical Installer"或"32-Bit Graphical Installer"下载安装程序(以 Windows 64 位操作系统为例)。

(2) 下载完成后，双击下载的文件，启动安装程序，进入 Anaconda 安装界面，如图 1-8 所示。

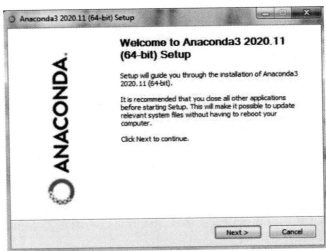

图 1-8　Anaconda 安装界面

如果在安装过程中遇到问题，可以暂时关闭杀毒软件后再次启动安装程序。

(3) 选择"Next"，进入 Anaconda 许可协议界面，如图 1-9 所示。

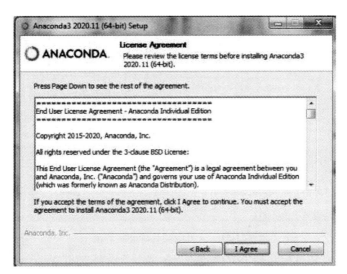

图 1-9　Anaconda 许可协议界面

(4) 阅读许可协议条款，勾选"I Agree"进入"Select Installation Type"(选择安装类型)界面，如图 1-10 所示。

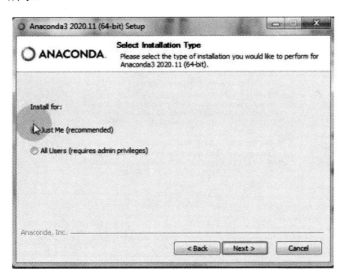

图 1-10　选择安装类型界面

如果在安装时选择了"All Users"(所有用户)项，应先卸载 Anaconda，然后选择"Just Me"(仅限本人)项重新安装。

(5) 除非是以管理员身份为所有用户安装，否则仅勾选"Just Me"并点击"Next"，进入"Choose Install Location"(选择安装目标路径)界面，如图 1-11 所示。

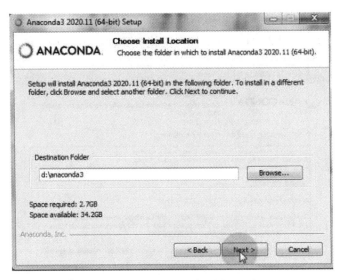

图 1-11　选择安装目标路径界面

(6) 在"选择安装目标路径"界面点击"Browse"选择安装 Anaconda 的目标路径,此处建议更改为默认安装路径,选择操作系统盘以外的其他磁盘根目录。目标路径中不能含有空格,也不能是"unicode"编码。然后点击"Next",进入"Advanced Installation Options"(高级安装选项)界面,如图 1-12 所示。

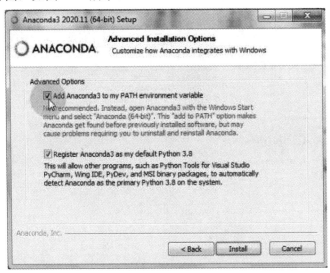

图 1-12　高级安装选项界面

在高级安装选项界面中,会看到"Add Anaconda3 to my PATH environment variable."(添加 Anaconda3 至我的环境变量)选项。如果在安装时勾选此项,安装完成后就无须配置环境变量,但可能会影响其他程序的使用。因此,建议在安装时不要勾选该选项,而是在安装完成后另行配置运行环境。

(7) 点击"Install"进行安装,安装过程中如果想要查看 Anaconda 安装细节,点击"Show details"即可,如图 1-13 所示。

第 1 章　Python 安装与管理

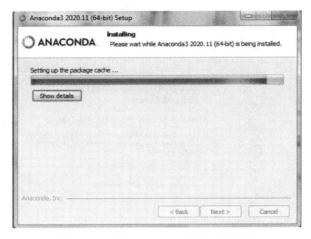

图 1-13　查看 Anaconda 安装细节界面

(8) 安装完成后显示"Installation Complete"(安装完成)界面，如图 1-14 所示。点击"Next"，出现如图 1-15 所示的 Anaconda 安装完成提示界面。

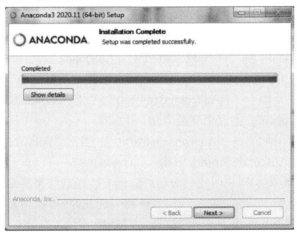

图 1-14　Anaconda 安装完成界面

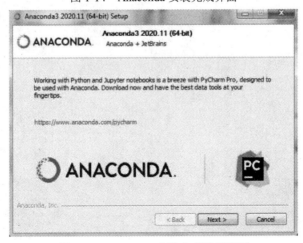

图 1-15　Anaconda 安装完成提示界面

(9) 点击图 1-15 中的"Next",进入"Thank you for installing Anaconda Individual Editon"安装成功界面,点击"Finish"完成安装,如图 1-16 所示。

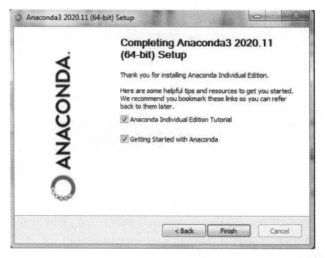

图 1-16 安装成功界面

如果不想观看 Anaconda 个人版学习教程而马上使用 Anaconda,则不勾选"Anaconda Individual Edition Tutorial"和"Getting Started with Anaconda"。

(10) 验证安装结果。可以选择以下任意一种方法验证是否安装成功。

① 点击计算机"开始"→"Anaconda3(64-bit)"→"Anaconda Navigator",若可以成功启动 Anaconda Navigator,则说明安装成功。

② 点击计算机"开始"→"Anaconda3(64-bit)",点击"Anaconda Prompt"→"以管理员身份运行",在 Anaconda Prompt 中输入"conda list",可以查看已经安装的包名和版本号。若能正常显示结果,则说明安装成功。验证安装成功结果界面如图 1-17 所示。

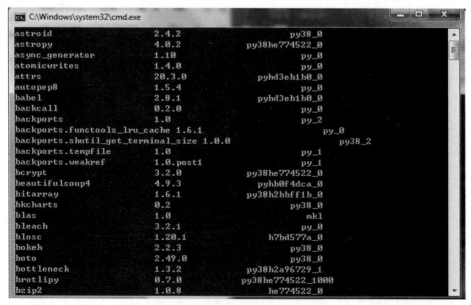

图 1-17 验证安装成功结果界面

③ 使用 Win + R 键入"cmd"后按"Enter"键，在提示符下输入"pip list"，查看已安装的内容，如图 1-18 所示。

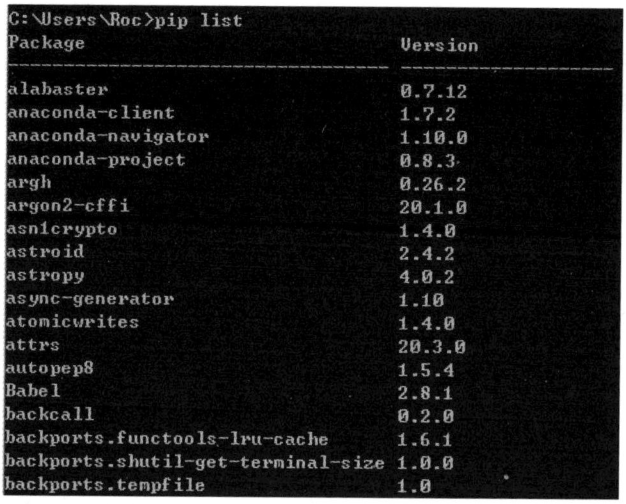

图 1-18 查看 Anaconda 安装包界面

2. Mac OS 系统安装 Anaconda

1) 图形界面安装

(1) 在 Anaconda 官方网站，根据个人计算机硬件条件选择安装文件，如点击"64-Bit Graphical Installer"进行下载。

(2) 完成下载之后，双击下载文件开始安装，在安装过程中"Introduction""Read Me"和"License"项可直接点击下一步。

(3) 在"Select a Destination"项的对话框中选择"Install for me only"并点击"Continue"，安装目标用户选择界面如图 1-19 所示。

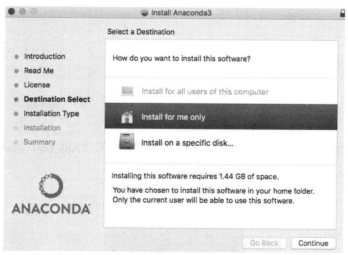

图 1-19 安装目标用户选择界面

若出现错误提示信息"You cannot install Anaconda in this location"，则重新选择"Install for me only"并点击"Continue"。

(4) 在"Installation Type"项的对话框中，点击"Change Install Location"来改变安装位置。若选择默认安装路径，可直接点击"Install"进行安装。选择安装位置的界面如图 1-20 所示。

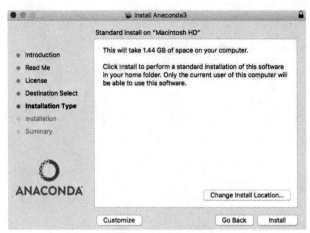

图 1-20　选择安装位置界面

(5) 程序安装完成后，在"Summary"项的界面中若看到"The installation was completed successfully"对话框则表示 Anaconda 安装成功，直接点击"Close"关闭对话框即可，如图 1-21 所示。

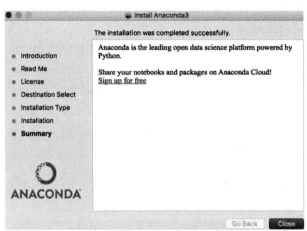

图 1-21　Anaconda 安装成功界面

(6) 安装成功后，在 Mac 的 Launchpad 中可以找到名为"Anaconda-Navigator"的图标，点击打开该图标启动程序，图标界面如图 1-22 所示。

图 1-22　Anaconda-Navigator 图标界面

(7) 若"Anaconda-Navigator"成功启动，则说明成功安装了 Anaconda，界面如图 1-23 所示；如果未成功，应仔细检查以上安装步骤重新安装。

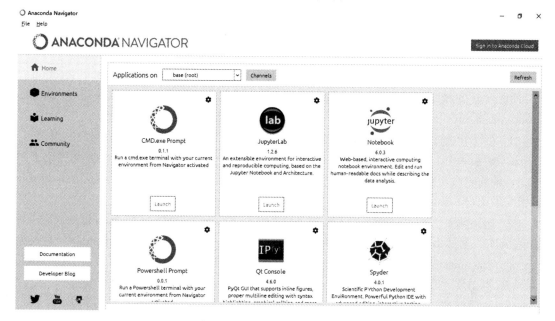

图 1-23 Anaconda-Navigator 界面

"Anaconda-Navigator"中包含"CMD.exe Prompt""JupyterLab""Notebook""PowerShell Prompt""Qt Console"和"Spyder"。

2) 命令行安装

(1) 在官方网站下载安装文件。选择 MacOX Python 3.8 版本后，根据硬件情况，选择相应的安装文件，如点击"64-Bit Command-Line Installer"下载。

(2) 下载完成之后，在 Mac 的 Launchpad 中找到"其他"并打开"终端"。以下安装步骤以 Python 3.6 为例。

在"终端"输入：

```
bash ~/Downloads/Anaconda3-5.0.1-MacOSX-x86_64.sh
```

注意：

① 无论是否用 bash shell，首词 bash 都需要输入。

② 如果下载路径是自定义的，那么把该步骤路径中的 ~/Downloads 替换成自己的下载路径。

③ 如果已将下载的.sh 文件重命名，那么将该步骤路径中的 Anaconda3-5.0.1-MacOSX-x86_64.sh 或 Anaconda2-5.0.1-MacOSX-x86_64.sh 替换成重命名后的文件名。

④ 尽量不要修改文件名。如果确实需要重命名，应使用英文进行命名。

(3) 在安装过程中，如果看到提示"In order to continue the installation process, please review the license agreement."(请浏览许可证协议以便继续安装)，点击"Enter"键查看"许可证协议"。

(4) 在"许可证协议"界面的底部输入"yes"，表示同意许可证协议内容，然后进行下一步。

(5) 在安装过程中，会提示“Press Enter to confirm the location, Press CTRL-C to cancel the installation or specify an alternate installation directory.”(按 Enter 键确认安装路径，按 CTRL-C 取消安装或者指定安装目录)。如果接受默认安装路径，则会显示“PREFIX=/home/<user>/anaconda<2 or 3>”并且继续安装。

(6) 安装时若提示“Do you wish the installer to prepend the Anaconda install location to PATH in your /home/<user>/.bash_profile?”(你希望安装程序添加 Anaconda 安装路径在/home/<user>/.bash_profile 文件中吗？)，建议输入“yes”。

① 路径 /home/<user>/.bash_profile 中的<user>是一个占位符，代表当前用户的用户名。在 macOS 系统中，用户的主目录通常位于 /Users/<user>下，因此 .bash_profile 文件位于用户主目录的隐藏文件夹中。例如，如果用户名是 wang，那么路径 /Users/wang/.bash_profile 就指向了用户 wang 的 .bash_profile 文件。

② 如果输入“no”，则需要手动添加路径。添加 export PATH=“/<path to anaconda>/bin:$PATH”到“.bashrc”或者“.bash_profile”中。其中，将“<path to anaconda>”替换为自己的 Anaconda 安装路径。

(7) 当出现“Thank you for installing Anaconda!”时，说明已经成功完成安装。

(8) 关闭终端，然后启动安装后的 Anaconda。

(9) 验证安装结果。可选用以下任意一种方法验证安装结果。

① 在终端输入命令“conda list”，如果 Anaconda 安装成功，则会显示已经安装的包名和版本号，如图 1-24 所示。

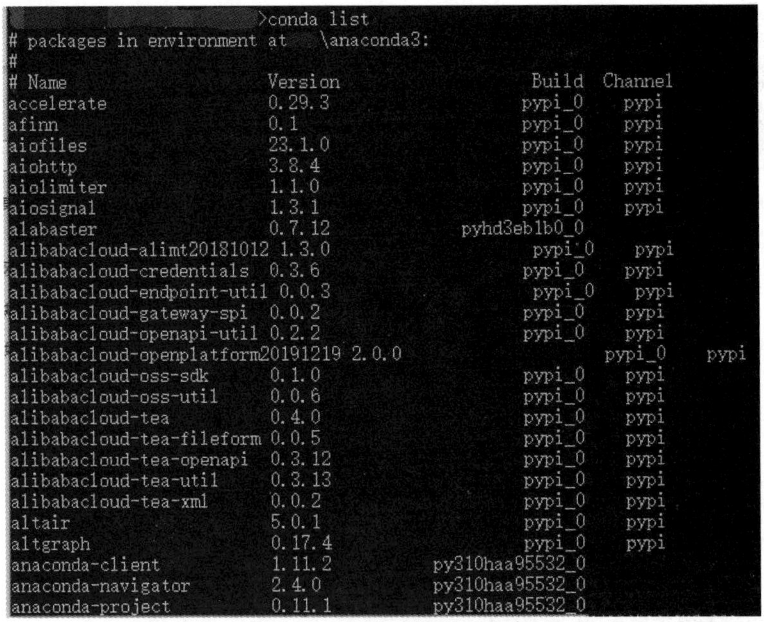

```
>conda list
# packages in environment at    \anaconda3:
#
# Name                    Version                   Build  Channel
accelerate                0.29.3                   pypi_0    pypi
afinn                     0.1                      pypi_0    pypi
aiofiles                  23.1.0                   pypi_0    pypi
aiohttp                   3.8.4                    pypi_0    pypi
aiolimiter                1.1.0                    pypi_0    pypi
aiosignal                 1.3.1                    pypi_0    pypi
alabaster                 0.7.12             pyhd3eb1b0_0
alibabacloud-alimt20181012 1.3.0                   pypi_0    pypi
alibabacloud-credentials  0.3.6                    pypi_0    pypi
alibabacloud-endpoint-util 0.0.3                   pypi_0    pypi
alibabacloud-gateway-spi  0.0.2                    pypi_0    pypi
alibabacloud-openapi-util 0.2.2                    pypi_0    pypi
alibabacloud-openplatform20191219 2.0.0                    pypi_0    pypi
alibabacloud-oss-sdk      0.1.0                    pypi_0    pypi
alibabacloud-oss-util     0.0.6                    pypi_0    pypi
alibabacloud-tea          0.4.0                    pypi_0    pypi
alibabacloud-tea-fileform 0.0.5                    pypi_0    pypi
alibabacloud-tea-openapi  0.3.12                   pypi_0    pypi
alibabacloud-tea-util     0.3.13                   pypi_0    pypi
alibabacloud-tea-xml      0.0.2                    pypi_0    pypi
altair                    5.0.1                    pypi_0    pypi
altgraph                  0.17.4                   pypi_0    pypi
anaconda-client           1.11.2          py310haa95532_0
anaconda-navigator        2.4.0           py310haa95532_0
anaconda-project          0.11.1          py310haa95532_0
```

图 1-24　Anaconda 已安装的包名和版本号界面

② 在终端输入“python”将会启动 Python 交互界面，如果成功安装 Anaconda 并且可以运行，则会在 Python 版本号的右边显示“Anaconda custom(64-bit)”，如图 1-25 所示。若

要退出 Python 交互界面，则输入"exit()"或"quit()"即可。

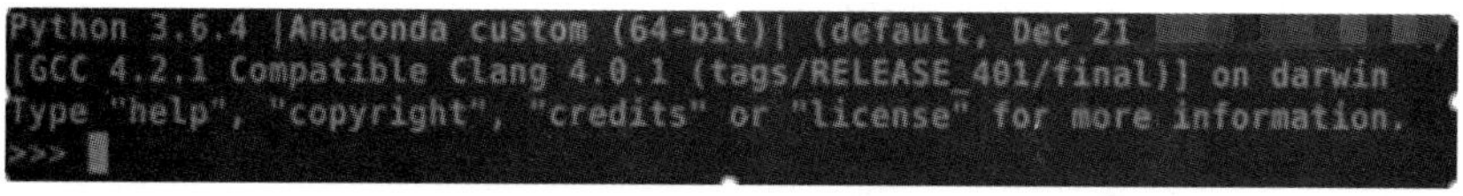

图 1-25 Python 交互界面

③ 在终端输入"anaconda-navigator"，并按"Enter"键。如果 Anaconda 成功安装，将会启动 Anaconda Navigator 的图形界面。

3. Linux 系统安装 Anaconda

(1) 在官方网站下载安装文件。

(2) 启动终端，在终端输入命令"md5sum/path/filename"或"sha256sum /path/filename"。

① 将该步骤命令中的"/path/filename"替换为文件的实际下载路径和文件名。其中，path 是路径，filename 为文件名。

② 路径和文件名中不能出现空格或其他特殊字符。

③ 路径和文件名应以英文命名，不应使用中文或其他特殊字符。

④ 根据 Python 版本，在终端输入命令(以 Python 3.6 为例)：

bash ~/Downloads/Anaconda3-5.0.1-Linux-x86_64.sh

⑤ 无论是否用 bash shell，首词 bash 都需要输入。

⑥ 如果下载路径是自定义的，那么把该步骤路径中的 ~/Downloads 替换成自己的下载路径。

⑦ 除非被要求使用 root 权限，否则均选择"Install Anaconda as a user"。

(3) 在安装过程中,会提示"In order to continue the installation process, please review the license agreement."(请浏览许可证协议以便继续安装)，可按"Enter"键查看"许可证协议"。

(4) 在"许可证协议"界面将屏幕滚动到底，输入"yes"表示同意许可证协议内容，然后进行下一步。

(5) 在安装过程中，会提示"Press Enter to accept the default install location, CTRL-C to cancel the installation or specify an alternate installation directory."(按 Enter 键确认安装路径，按 CTRL-C 取消安装或者指定安装目录)。如果接受默认安装路径，则会显示"PREFIX=/home/<user>/anaconda<2 or 3>"并且继续安装。

(6) 安装时若提示"Do you wish the installer to prepend the Anaconda <2 or 3> install location to PATH in your /home/<user>/.bashrc?"(你希望安装程序添加 Anaconda 安装路径在/home/<user>/.bashrc 文件中吗？)建议输入"yes"。

(7) 当出现"Thank you for installing Anaconda <2 or 3>!"时，说明已经成功完成安装。

(8) 关闭终端，然后启动安装后的 Anaconda，或者直接在终端中输入"source ~/.bashrc"完成启动。

(9) 验证安装结果。可选用以下任意一种方法验证是否安装成功。

① 在终端输入命令"condal list"，如果 Anaconda 安装成功，则显示已经安装的包名和版本号。

② 在终端输入"python"，启动 Python 交互界面。若 Anaconda 安装成功并且可以运

行，则会在 Python 版本号的右边显示"Anaconda custom(64-bit)"。若要退出 Python 交互界面，则输入"exit()"或"quit()"即可。

③ 在终端输入"anaconda-navigator"，若 Anaconda 成功安装，则会启动 Anaconda Navigator。

1.2 环境变量配置

1. 文件路径

若在 Windows 系统下安装 Anaconda 时未勾选"Add Anaconda to my PATH environment variable"，则需要配置环境变量。以 Anaconda 装在 D 盘为例，需要把以下路径添加到系统环境变量中：

D:\Anaconda3;

D:\Anaconda3\Scripts;

D:\Anaconda3\Library\mingw-w64\bin;

D:\Anaconda3\Library\usr\bin;

D:\Anaconda3\Library\bin;

2. 配置方法

(1) 点击"开始"，找到"控制面板"，选择"系统与安全"，打开"系统"，显示高级系统设置界面，如图 1-26 所示。

图 1-26　高级系统设置界面

(2) 点击"高级系统设置"，打开如图 1-27 所示的系统属性界面。也可以使用鼠标右键点击"我的电脑"，选择"属性"，然后点击"高级系统设置"。

第 1 章　Python 安装与管理

图 1-27　系统属性界面

(3) 点击"环境变量"后进入编辑系统变量界面，如图 1-28 所示。

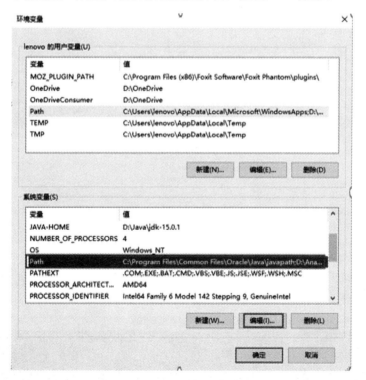

图 1-28　编辑系统变量界面

(4) 点击"编辑"后开始编辑环境变量，如图 1-29 所示。

图 1-29　编辑环境变量

(5) 点击"新建",把路径名称依次复制到空白对话框里,点击"确定",如图 1-30 所示。

图 1-30　添加环境变量界面

(6) 检查 Conda。在命令提示符窗口中输入"conda --version",界面如图 1-31 所示。若输出版本信息,则表示配置环境变量成功。

图 1-31　检查 Conda 环境变量配置界面

如果出现错误信息，则需核实是否出现以下情况：当前账户是不是当时安装 Anaconda 的账户；安装 Anaconda 之后是否重启了电脑。

1.3　Conda 运行环境及管理

Conda 是一个开源的软件包管理系统和环境管理系统，用于科学计算、数据分析和机器学习等领域。它最初是为 Python 开发的，但也能管理和安装非 Python 软件包。Conda 能够帮助用户快速安装、升级和管理软件包，解决了依赖关系管理的问题。用户可以使用 Conda 来创建多个独立的环境，每个环境都有自己的软件包集合，这样可以避免不同软件包之间的冲突。这对于开发和部署项目非常有用，因为可以为每个项目创建一个独立的环境，确保项目所需的依赖关系不会与其他项目发生冲突。Conda 还提供了一个广泛的软件包仓库，用户可以从中获取各种各样的软件包，包括科学计算、数据处理、机器学习和深度学习工具等。此外，Conda 还允许用户创建自己的软件包，并与其他用户共享。总之，Conda 是一个强大的工具，可以帮助用户轻松管理软件包和环境，提高开发效率，同时保持系统的整洁和稳定。

Conda 运行管理的各个命令均在命令行模式执行，Windows 用户打开"Anaconda Prompt"或按 Win + R 键，输入"cmd"后按"Enter"键进入 DOS 命令行模式进行操作；MacOS 和 Linux 用户则应打开"Terminal"进行操作。本节内容中"env_name"是创建的环境名，应用英文命名，且名字中间不加空格。"package_name"是单个包名，"package_names"是多个包名。环境名和包名前后用空格隔开，包名前后用空格隔开。

1. Conda 运行环境

1) 创建环境

创建环境命令格式如下：

```
conda create --name env_name package_names
```

例如，要创建 Python 3.8 版本运行环境，应在 DOS 命令行窗口输入以下命令：

```
conda create -n python38 python=3.8
```

或

```
conda create --name python38 python=3.8
```

在上面的命令中，"python38"是设置环境的名称，"-n"是指该命令后面的"python38"是要创建环境的名称。

如果要在新创建的环境中创建多个包，则直接在"package_names"后添加多个包名即可，包与包之间用空格隔开。例如，"conda create -n python38 python=3.8 numpy pandas"，即创建一个名为"python38"的环境，环境中安装了版本为 3.8 的 Python，同时也安装了 Numpy 和 Pandas。注："name"同样可以替换为"-n"。

在默认情况下，新创建的环境将会被保存在"/Users/<user_name>/anaconda3/env"目录下。其中，"<user_name>"为当前电脑用户的用户名。在创建环境时，可以指定即将安装在该环境中的 Python 版本。当同时使用 Python 2.x 和 Python 3.x 中的代码时，这种操作很有用。

2) 进入创建的环境

进入创建的环境时要在 DOS 命令行窗口输入以下命令：

conda activate python38

进入之后，可以在终端提示符中看到环境名称"python38"。进入环境之后，可以用"conda list"或"pip list"命令查看环境中默认的安装包。

3) 离开环境

要离开当前环境，应在 DOS 命令行窗口输入以下命令：

deactivate

4) 共享环境

共享环境的作用是让其他人共享代码中使用的所有包，并确保这些包的版本正确。比如一个新开发的系统，在进行项目部署时，部署人并不知道系统在开发时使用的是哪个 Python 版本，以及使用了哪些包及版本情况。此时在当前的环境终端中可使用以下命令将当前环境保存到后缀名为 YAML 的文件(包括 Pyhton 版本和所有包的名称)中：

conda env export > environment.yaml

命令的第一部分"conda env export"用于输出环境中所有包的名称(包括 Python 版本)。在终端上可以看到导出的环境文件路径。在 GitHub 上共享代码时，最好同样创建环境文件并将其包括在代码库中，这能让其他人更轻松地安装共享代码的所有依赖项。

5) 导出的环境文件

使用导出的环境文件时，首先在 Conda 中进入环境，比如"conda activate python38"，然后使用以下命令更新环境：

conda env update -f=/path/to/environment.yml

其中，"-f"表示要导出文件在本地的路径，所以"/path/to/environment.yml"要换成本地的实际路径。

对于不使用 Conda 的用户，可以使用以下命令将 txt 文件导出并包括在其中：

pip freeze > environment.txt

将该文件包含在项目的代码库中，其他电脑上即使没有安装 Conda 也可以使用该文件安装一样的开发环境。使用者在自己的电脑上进入 Python 命令环境，然后运行以下命令就可以安装该项目需要的包：

pip install -r C:\Users\Microstrong\enviroment.txt

其中，"C:\Users\Microstrong\enviroment.txt"是该文件在运行电脑上的实际路径。

6) 列出环境

有时使用者会忘记自己创建的环境名称，这时用 conda env list 就可以列出自己创建的所有环境，当前所在环境的旁边会有一个星号标识。默认的环境(即不在选定环境中时使用的环境)名为"base"，也可以使用以下命令查看环境：

conda info --envs

或

conda info -e

查看已创建的环境界面如图 1-32 所示。

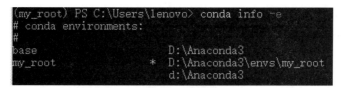

图 1-32　查看已创建的环境界面

7) 删除环境

如果不再使用某个环境，可以使用以下命令删除环境：

conda env remove -n python38

这里环境名为"python38"。

8) 更新 Conda

更新 Conda 至最新版本，可以使用以下命令：

conda update conda

执行命令后，Conda 将会对版本进行比较并列出可以升级的版本。同时告知用户其他相关包也会升级到相应版本。当较新的版本可以用于升级时，终端会显示"proceed (y/n)?"此时输入"y"即可进行升级。

9) 查看 Conda 帮助信息

查看 Conda 帮助信息的命令如下：

conda --help

或

conda -h

10) 卸载 Conda

(1) 在 Linux 或 macOS 中卸载。在 Linux 或 macOS 中卸载 Conda 的命令如下：

rm -rf ~/anaconda2

或

rm -rf ~/anaconda3

根据安装的 Anaconda 版本选择相应的卸载命令即可删除 Anaconda 的安装目录。

(2) 在 Windows 系统中卸载。在 Windows 系统中卸载 Conda 的步骤是：

选择"控制面板"→"添加或删除程序"→"Python X.X (Anaconda)"→"删除程序"。其中，"Python X.X"即 Python 的版本，如 Python 3.6，3.7 或 3.8 等。

Windows 10 系统下的 Conda 卸载方式有所不同，具体方法如下：

点击"开始"→"控制面板"→"添加或删除"→"程序"→"Anaconda 3"(Python X.X)，将光标移到程序，点击鼠标右键，选择"卸载"。

2. 管理 Anaconda 安装包

安装 Anaconda 之后，可以很方便地安装、卸载或更新安装包。

1) 安装包

(1) 在指定环境中安装包。在指定环境中安装包的命令如下：

conda install --name env_name package_name

例如，conda install --name python2 pandas 即在名为"python2"的环境中安装 Pandas 包。

(2) 在当前环境中安装包。在当前环境中安装包的命令如下：

```
conda install package_name
```

例如，conda install pandas 即在当前环境中安装 Pandas 包。

(3) 使用 pip 安装包。

当使用 conda install 无法进行安装时，可以使用 pip 进行安装。使用 pip 安装包的命令如下：

```
pip install package_name
```

例如，pip install driver 即安装 driver 包。

注意：

① pip 只是包管理器，无法对环境进行管理。因此如果想在指定环境中使用 pip 安装包，则需要先切换到指定环境中再使用 pip 命令。

② pip 无法更新 Python，因为 pip 并不将 Python 视为包。

③ pip 可以安装一些 Conda 无法安装的包，Conda 也可以安装一些 pip 无法安装的包。因此当使用一种命令无法安装包时，可以尝试使用另一种命令。

(4) 从 Anaconda.org 安装包。

当使用 conda install 无法进行安装时，可以考虑从 Anaconda.org 中获取安装包的命令并进行安装。从 Anaconda.org 安装包时，无须注册。在当前环境中安装来自 Anaconda.org 的包时，需要指定包所在的途径(channel)。

查询路径和安装方法如下：

① 在浏览器中输入"http://anaconda.org"，或直接点击"Anaconda.org"。

② 在新页面"Anaconda Cloud"的搜索框中输入要安装的包名，然后点击右边的放大镜标志，界面如图 1-33 所示。

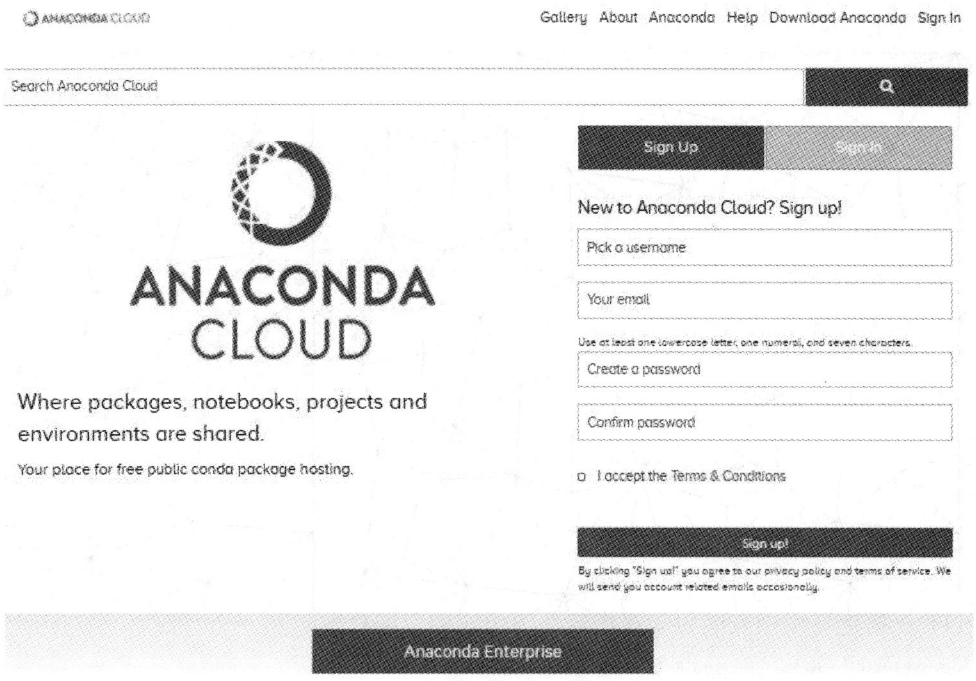

图 1-33　Anaconda Cloud 搜索界面

③ 搜索结果中有数以千计的包可供选择，点击"Downloads"可根据下载量进行排序，排在最上面的是下载量最多的包，如图 1-34 所示。

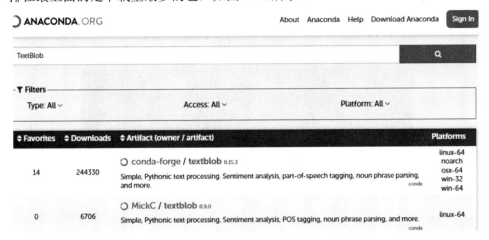

图 1-34　搜索 TextBlob 界面

④ 选择满足需求的包，点击包名。

⑤ 复制"To install this package with conda run:"下方的命令，并粘贴在终端，执行后即完成安装。

2) 卸载包

(1) 卸载指定环境中的包，使用以下命令：

conda remove --name env_name package_name

例如，conda remove --name python3 numpy 即卸载名为"python3"中的 numpy 包。

(2) 卸载当前环境中的包，应使用以下命令：

conda remove package_name

例如，conda remove matplotlib 即在当前环境中卸载 matplotlib 包。

3) 更新包

(1) 更新所有包。更新全部包的命令如下：

conda update --all

或

conda upgrade --all

建议在安装 Anaconda 之后执行上述命令更新 Anaconda 中的所有包至最新版本，以便使用。

(2) 更新指定包。更新指定包的步骤如下：

① 使用以下命令查看待更新的包：

pip list --outdated

conda update package_name

或

conda upgrade package_name

例如，要更新 matplotlib，在 DOS 命令行窗口输入以下命令：

conda update matplotlib

或

conda upgrade matplotlib

使用 conda update python 更新 Python，如图 1-35 所示。

图 1-35　更新 Python

更新多个指定包时，用空格隔开包名，向后排列。例如，conda update pandas numpy matplotlib 即更新 pandas、numpy、matplotlib 包。

② 使用以下命令安装更新工具 pip-review：

pip install pip-review

执行下面的代码后，将逐个弹出窗口询问某个库是否需要更新：

pip-review --local –interactive

③ Python 脚本程序自动更新，示例代码如下：

import pip

from subprocess import call

from pip._internal.utils.misc import get_installed_distributions

for dist in get_installed_distributions():

　　call("pip install --upgrade " + dist.project_name, shell=True)

4）查找可供安装的包版本

(1) 精确查找使用以下命令：

conda search --full-name <package_full_name>

其中，--full-name 为精确查找的参数，package_full_name 是被查找包的全名。

例如，conda search --full-name python 即查找全名为"python"的包有哪些版本可供安装。

(2) 模糊查找使用以下命令：

conda search text

其中，text 是查找含有此字段的包名。

例如，conda search py 即查找含有"py"字段的包有哪些版本可供安装。

使用以上命令时，当系统报错"conda 不是系统内部命令"时，可参照配环境变量的方法解决。

第 2 章 Python 编辑器

2.1 常用编辑器概况

本章主要介绍常用 Python 编辑器的概况，并选择 Jupyter、PyCharm、NotePad 和 Sublime Text 四种编辑器示范 Python 编辑器的安装和使用。推荐初学者使用 Sublime Text 编辑器，它具有体积小、安装便捷、占内存小、使用方便等优点。初学者也可以使用 Python 自带的编辑器 IDLE。以下简要介绍一些常见编辑器的特点。

(1) IDLE：简单通用，支持多种设备，可用于 Python 基础学习；具有内置修正、突出显示错误、自动编码等功能，用户可在该编辑器中进行搜索、替换操作。安装 Python 程序后可在"开始"菜单中找到"IDLE"选项，双击该选项即可运行程序。

(2) Jupyter/IPython Notebook：基于 Web 的编辑器，支持运行 40 多种编程语言，相对简单，对用户友好，可用于数据可视化。

(3) PyCharm：有一整套可以提高 Python 语言开发效率的工具，比如调试、语法高亮、Project 管理、代码跳转、智能提示、自动完成、单元测试、版本控制。此外，该编辑器还提供了一些高级功能，用于支持 Django 等框架下的专业 Web 开发，适合专业程序开发人员使用；该编辑器界面较复杂，不建议新手使用。

(4) Sublime Text：一款主流的前端开发编辑器，体积较小，运行速度快，文本功能强大；支持编译功能且可在控制台看到输出；内嵌的 Python 解释器可支持插件开发；支持大量插件，并可对其进行管理。Sublime Text 是一个跨平台的编辑器，同时支持 Windows、Linux、MacOS 等操作系统。

(5) NotePad：免费开源，速度快，容易使用，对用户友好，支持正则表达式，具有多窗口显示、缩放、录制回放功能，所见即所得；配置后支持 Java、HTML、Pascal、perl、php、Python 等多种语言的语法。

(6) UltraEdit：非开源文本编辑器，占用内存小，速度较快，兼容 Windows、Mac 和 Linux 系统，内置代码支持 HTML、PHP 和 JavaScript 等语法，配置后支持 Python 解释器，有代码提示、折叠功能，配置信息和功能较全面，右键菜单功能较多。

(7) Thonny：界面简单，适合新手程序员使用，每一步都有对表达式的评估。

(8) Pyzo：免费开源编辑器，可与其他 Python 解释器使用，具有良好的简洁性和交互性，使用简单，很适合新手使用。

(9) Visual Studio Code：可智能感知错误，误报错率较低，可断点调试，有显示、编辑器、文件管理、多行光标、导航、调试、集成终端等多种快捷键；有中文菜单模式，便于新手学习；可通过加载插件和配置用户设置等来添加新语言、主题、调试器、附加服务。

(10) VI/Vim：兼容多平台，支持多个插件管理版本、文件等，可突出显示搜索结果，其脚本允许 Python 执行大部分编程任务，位列最好用 Python IDE 工具前五名。

(11) GNU Emacs：编辑器可扩展、实时显示、自动文档化、兼容性高，支持 Unicode 编码，提供大量自定义脚本供开发者使用。

(12) Atom：是一款开源编辑器，可与 Java、Php 等大部分编程语言兼容，含多个插件，功能强大。

(13) Spyder：主要用于数据科学，是开源工具，兼容多平台，合并多个库，如 Matplotlib、Numpy 等，IDE 新手也可使用。

(14) Eclipse + PyDev：Windows 首选的开源编辑器之一，Eclipse 与 PyDev 扩展项结合可实现智能修复错误。

(15) Eric Python：一款功能强大、高效的编辑器，可在多平台运行。

(16) Wing：简单易用、修复功能强大、可实现智能编程，可远程编程。

(17) PyScripter：轻量级编辑器，在 Mac 系统使用较多，功能丰富，其商业版支持大部分编程语言。

2.2　Jupyter 安装与使用

1. Jupyter Notebook 概述

Jupyter Notebook 是基于浏览器的编辑器，系统成功安装 Anaconda 后便带有此编辑器，可在电脑"开始"→"所有程序"中双击此程序图标运行后使用。也可在官方网站下载安装后使用。它的优点是可以在网页上分段(按单元格)展示 Python 代码的运行结果，便于调试和保存中间结果。

2. Jupyter 的使用方法

点击电脑"开始"，打开"所有程序"，双击"Jupyter Notebook"运行程序界面，如图 2-1 所示。

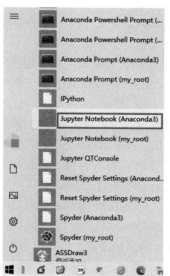

图 2-1　启动 Jupyter Notebook 界面

第 2 章　Python 编辑器

在 Jupyter 界面选择右上的"New",新建 Python 文件,界面如图 2-2 所示。

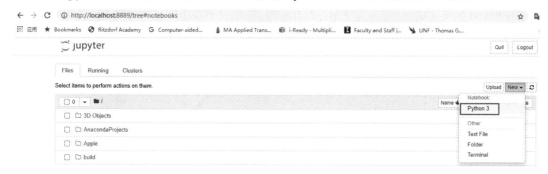

图 2-2　新建 Python 程序文件界面

在新弹出窗口中的绿色方框处输入 Python 代码,按"Ctrl"+"Enter"即可得到运行结果,结果会显示在单元格的下方,如图 2-3 所示。

图 2-3　Python 输出运行结果界面

2.3　NotePad 安装与使用

1. NotePad 的安装

NotePad 是一款免费、开源的文本、源码编辑器。其下载界面如图 2-4 所示。

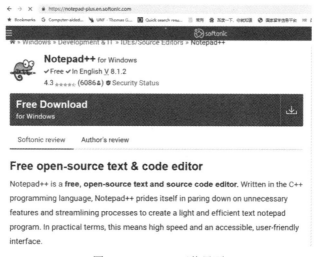

图 2-4　NotePad 下载界面

2. NotePad 的配置与使用

NotePad 安装完成后，双击程序图标打开程序，按照以下步骤配置 NotePad(以汉化版为例)。

(1) NotePad 启动后界面如图 2-5 所示，依次单击图中①②项。

图 2-5　NotePad 启动后界面

(2) 选择 Python 程序所在位置后保存，如图 2-6 所示(图中 Python 程序位于 E 盘 anaconda3 文件夹内)。

图 2-6　保存 Python 程序界面

(3) 创建快捷方式界面如图 2-7 所示，标记①处为命名快捷方式的名称，标号记②或③，和④和⑤处为选择合适的选项创建快捷方式，选择完成后点击"OK"。

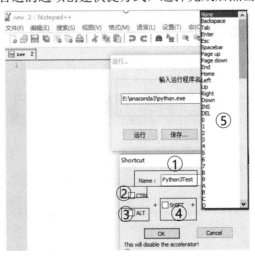

图 2-7　创建快捷方式界面

(4) 点击菜单"运行(R)",选择新建的快捷方式"Python3Test"后双击运行 Python 程序,运行 NotePad 界面如图 2-8 所示。

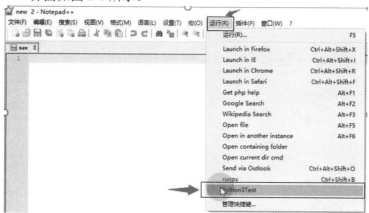

图 2-8　运行 NotePad 界面

(5) 在提示符下输入 Python 程序,按"Enter"键后运行程序。若程序可以正常运行,表明完成配置 NotePad,NotePad 编辑器运行 Python 界面如图 2-9 所示。

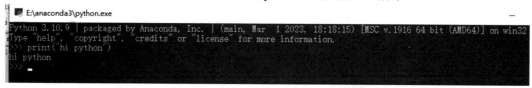

图 2-9　NotePad 编辑器运行 Python 界面

2.4　Sublime Text 的配置与使用

1. Sublime Textd 的下载与配置

Sublime Text 的用户界面友好、功能强大,有代码缩略图、Python 插件、代码段等。可设置自定义键,菜单和工具栏。主要有拼写检查、书签、完整的 Python API、Goto、即时项目切换、多选择、多窗口等功能。常见功能包括自动完成、多列编辑、代码注释、行操作和快捷键等。Sublime Text 下载页面如图 2-10 所示。

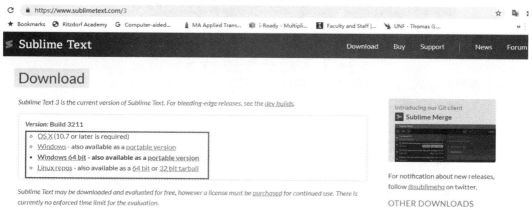

图 2-10　Sublime Text 下载页面

Sublime Text 的下载与配置步骤如下。

(1) 根据不同系统选择不同平台的版本，此处选择下载 Windows 64 位系统免安装版 Sublime Text 并解压缩，解压缩后的界面如图 2-11 所示。

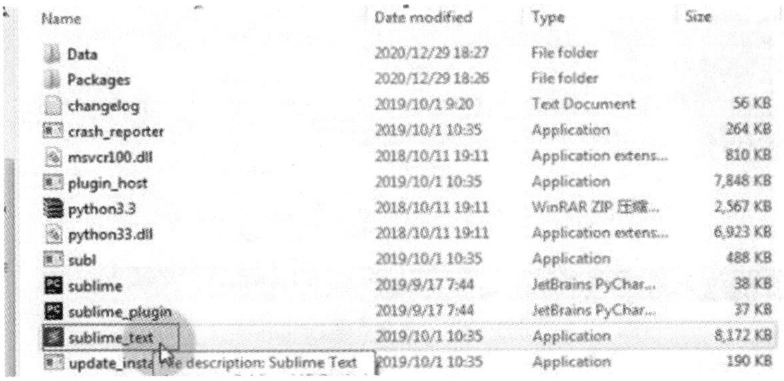

图 2-11　Sublime Text 解压缩后的界面

(2) 双击运行程序后，依次选择对话框中标记①(Tools)、②(Build System)和③(New Build System)处新建 Sublime Text 系统，界面如图 2-12 所示。

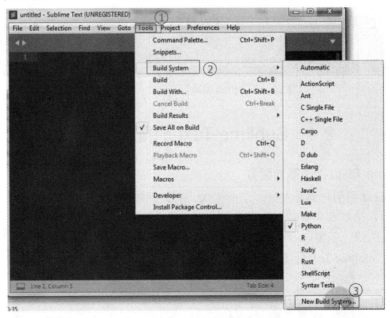

图 2-12　新建 Sublime Text 系统界面

(3) 新建 Sublime Text 系统默认配置界面如图 2-13 所示。

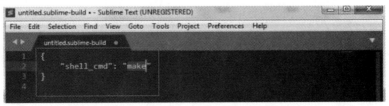

图 2-13　Sublime Text 系统默认配置界面

(4) 为了使系统支持中文，需使用图 2-14 中代码替换图 2-13 中的代码。代码中"D:/Anaconda3/python.exe"是 Python 程序所在的路径，配置时根据自己的 Python 程序路径更改。

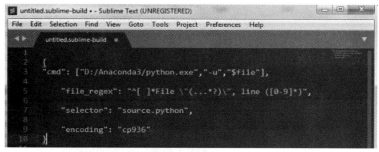

图 2-14　配置 Sublime Text 系统支持中文界面

(5) 命名新建系统名称并保存，如图 2-15 所示。

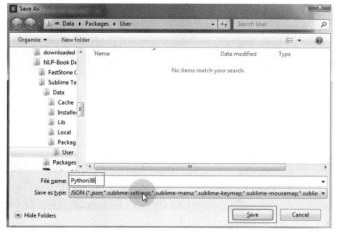

图 2-15　命名并保存新建系统界面

(6) 点击菜单"Tools"，在"Build System"菜单里勾选红框所示新建系统 Python38，完成新系统创建，如图 2-16 所示。

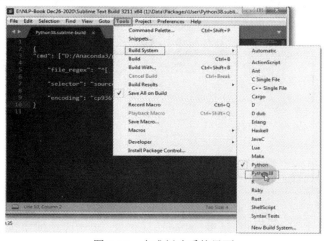

图 2-16　完成新建系统界面

2. Sublime Text 的配置与使用

在新系统创建完成后即可使用，先新建一个文件，再写入代码后保存为 Python 文件，测试运行结果。具体使用与配置的步骤如下。

(1) 点击"File"，选择"New File"，新建一个 Python 文件，如图 2-17 所示。

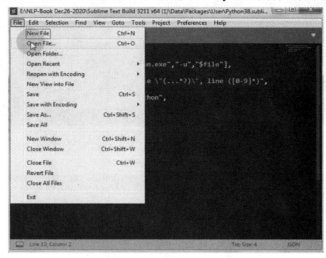

图 2-17　新建 Python 文件界面

(2) 在指定位置写入代码，如图 2-18 所示。

图 2-18　写入 Python 代码界面

(3) 在"File"菜单里选择"Save As…"选择保存文件保存位置，命名文件，选择 Python 文件格式并保存，如图 2-19 所示。

图 2-19　保存 Python 文件界面

(4) 执行代码，测试代码运行结果。代码运行正常，表明成功创建新系统，Sublime Text 配置完成。运行结果如图 2-20 所示。

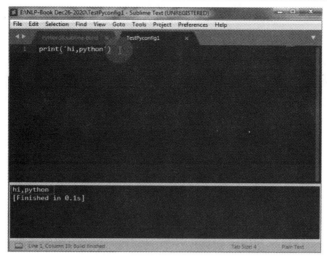

图 2-20　测试代码运行结果界面

3. Sublime Text 常用快捷键

Sublime Text 常用快捷键功能如表 2-1 所示。

表 2-1　Sublime Text 常用快捷键功能

快 捷 键	功　　能
Ctrl + N	新建
Ctrl + B	执行命令
Ctrl + S	保存
Ctrl + A	选定全部
Ctrl + /	注释
Ctrl + Shift + /	当前位置插入注释
Ctrl + 鼠标滚动向上/向下	字体变大/缩小
Ctrl + =/-	字体变大/缩小
Ctrl + Shift + k	删除一行
Alt +	快速闭合标签
Ctrl + P	打开搜索对话框，输入搜索任意文本字符
Ctrl + 回车	添加一行空行
Ctrl + Shift + V	粘贴过程中保持缩进
Alt + F3	选中选择的词
Ctrl + W	关闭当前文档
Ctrl + F	查找
Ctrl + H	查找 + 替换
Ctrl + D	选中光标所在的文本，继续操作则会选中下一个相同的文本
Ctrl + Shift + D	复制光标所在整行插入到下一行
Shift + Tab	去除缩进

4. 部分 Sublime Text 的常用设置

(1) 自动开启换行设置。选择菜单"View"→"Word Wrap"。

(2) 单行注释用"#"放在行首，例如

#print('hello, world')

表示这行是注释，不会被当作程序执行。

(3) 多行注释用三个单引号或三个双引号，例如

'''
AAA
BBB
CCC
'''

表示注释三个单引号或三个双引号里面的内容，执行程序时注释内容不会被当作命令执行。

2.5 PyCharm 的安装与使用

1. PyCharm 概述

PyCharm 可以在 Windows、MaxOS 和 Linux 系统使用，它支持 Web 开发框架，如 Paramid、Flask 和 Django；提供智能代码功能，可快速、精确地修复程序中的错误。用 Doker 和 Vagrant 连接，并在 ssh 终端帮助下，可在远程主机上进行程序开发。PyCharm 官网有三个版本，分别是专业版、社区版和教育版。专业版功能更加强大，主要是为 Python 和 Web 开发者而准备，支持远程开发，具有数据库工具、Web 开发工具等高级功能，需要付费使用。社区版是免费开源的版本，提供基本的 Python 开发功能。教育版通常提供给授权的教育机构和学生免费使用，更侧重于教学和学习，提供了一些额外的功能和工具，以支持教学环境中的教学和学习活动。

2. PyCharm 社区版的安装和使用

1) 在 Windows 系统安装 PyCharm 社区版的步骤

(1) 在官方网站下载 PyCharm 社区版安装文件，下载界面如图 2-21 所示。

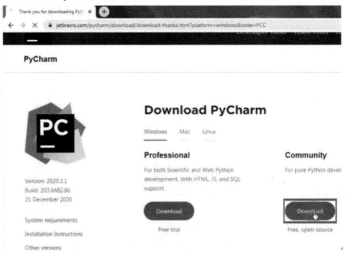

图 2-21　PyCharm 下载界面

(2) 下载完成后,双击 PyCharm 程序图标安装。安装界面如图 2-22 所示,点击"Next"安装。

图 2-22　安装 PyCharm 程序初始界面

(3) 点击"Browse"选择安装文件路径,如图 2-23 所示。点击"Next"进入下一步。

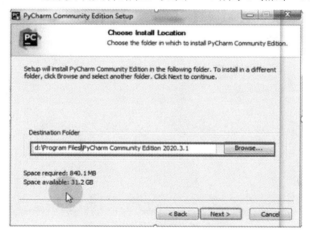

图 2-23　选择安装文件位置界面

(4) 创建桌面快捷方式和关联 py 文件界面如图 2-24 所示,勾选①、②处,点击"Next"进入下一步。

图 2-24　创建桌面快捷方式并关联 py 文件界面

(5) 点击"Install"开始安装 PyCharm，界面如图 2-25 所示。

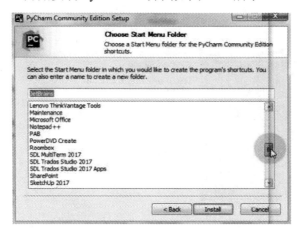

图 2-25　开始安装 PyCharm 界面

(6) 点击"Show details"查看安装细节，等待安装完成，如图 2-26 所示。

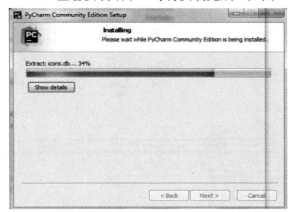

图 2-26　查看安装细节界面

(7) 点击"Finish"完成安装，如图 2-27 所示；若需运行 PyCharm，可先勾选"Run PyCharm Community Edition"。

图 2-27　完成安装界面

2) PyCharm 社区版的使用步骤

完成安装后，若已勾选 "Run PyCharm Community Edition"，PyCharm 会自动运行；若未勾选，可以在桌面双击 PyCharm 图标运行程序。

(1) 程序运行界面如图 2-28 所示，勾选图中标①处确认已读并接受用户协议。

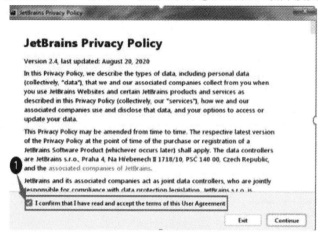

图 2-28　程序运行界面

(2) 在数据共享界面选择不发送或匿名发送数据，如图 2-29 所示。

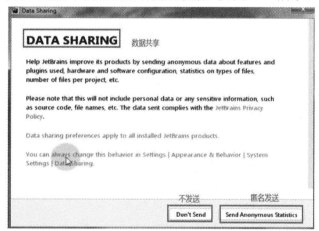

图 2-29　数据共享界面

(3) 新建项目界面如图 2-30 所示，点击新建项目设置项目路径、关联 Python 程序。

图 2-30　新建项目界面

(4) 设置项目路径、关联 Python 程序界面，如图 2-31 所示。点击图中标记①处的文件夹图标更改项目位置，图中项目处于 H 盘 PythonProject 文件夹中；标记②处文件夹路径会自动生成，不用更改；标记③处是默认的 Python 程序文件夹，点击图中标记④处"Create"后会自动下载 Python 程序。如果已安装 Anaconda，点击标记③处文件夹图标选择 Anaconda 所在文件夹，选择文件夹中 Python 程序路径后点击"Create"创建项目。例如，Anaconda3 安装在 E 盘根目录下，路径如图中标记③处所示。

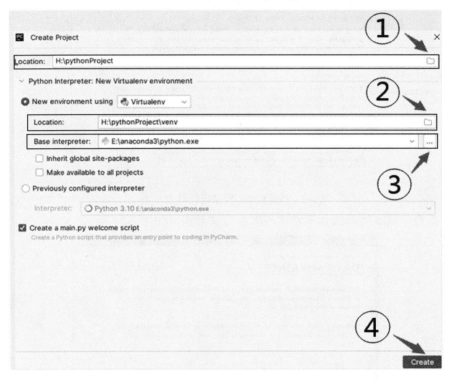

图 2-31　设置项目路径、关联 Python 程序界面

(5) 创建 Python 程序的界面如图 2-32 所示。点击图中左上角"File"，选择菜单中"New"，选择图中"Python File"创建 Python 程序。

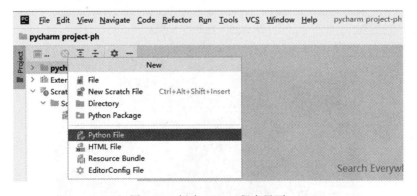

图 2-32　创建 Python 程序界面

(6) 输入 Python 语句并运行程序，界面如图 2-33 所示，点击图中①处输入 Python 语

句，点击②处运行程序，运行结果如③处所示。

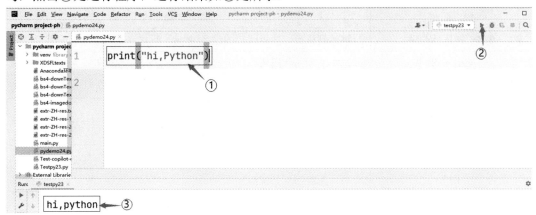

图 2-33　输入 Python 语句并运行程序界面

3. PyCharm 的常用快捷键

PyCharm 常用快捷键和功能如表 2-2 所示。

表 2-2　PyCharm 常用快捷键及功能

快 捷 键	功　　能
Ctrl + Q	快速查看文档
Ctrl + F1	显示错误描述或警告信息
Ctrl + /	行注释(可选中多行)
Ctrl + Alt + L	代码格式化
Ctrl + Alt + O	自动导入
Ctrl + Alt + I	自动缩进
Tab/Shift + Tab	缩进、不缩进当前行(可选中多行)
Ctrl + C/Ctrl + Insert	复制当前行或选定的代码块到剪贴板
Ctrl + D	复制选定的区域
Ctrl + Y	删除当前行
Shift + Enter	换行
Ctrl + J	插入模版
Ctrl + Shift + /-	展开/折叠全部代码块
Ctrl + Numpad +	全部展开
Ctrl + Numpad -	全部折叠
Ctrl + Delete	删除到字符结束
Ctrl + Backspace	删除到字符开始
Ctrl + Shift + F7	将当前单词在整个文件中高亮，F3 移动到下一个，Esc 取消高亮
Alt + up/down	跳转到上一个/下一个方法
Alt + Shift + up/down	当前行上移或下移

续表

快 捷 键	功　能
Ctrl + B/鼠标左键	跳转到函数、变量或类的定义/声明
Ctrl + W	选中增加的代码块
Shift + F6	重命名方法或变量
Ctrl + E	最近访问的文件
Esc	从其他窗口回到编辑窗口
Shift + Esc	隐藏当前窗口
F12	回到先前的工具窗口
Ctrl + Shift + up	快速上移到某一行
Ctrl + Shift + down	快速下移到某一行
Ctrl + Alt + 左箭头	返回上一个光标的位置(Ctrl 进入函数后返回)
Ctrl + Alt + 右箭头	前进到后一个光标的位置

第 3 章 Python 基础

3.1 Python 常识

1. Python 的输入与输出

1) input()输入

在 Python3 中，使用键盘输入 input()获取数据，获取到的数据会存放到等号左边的变量中；但在 Python2 中，使用的是 raw_input()。input() 函数用于向用户生成一条提示，获取用户输入的内容。由于 input()函数总会将用户输入的内容放入字符串中，因此用户可以输入任何内容，input()函数总是返回字符串。例如：

msg1 = input("请输入你的信息：")

运行程序输出结果如下：

请输入你的信息：

print(type(msg1))

print(msg1)

第一次运行该程序，我们输入一个整数 3，运行过程如下：

请输入你的值：3

\<class 'str'>

3

第二次运行该程序，我们输入一个浮点数 0.7，运行过程如下：

请输入你的值：0.7

\<class 'str'>

0.7

第三次运行该程序，我们输入一个字符串 hi, python，运行过程如下：

请输入你的值：hi,python

\<class 'str'>

hi,python

无论输入哪种内容，input()函数始终返回字符串，程序总会将用户输入的内容转换成字符串。

2) 格式化输出

格式化输出是在输出代码中使用操作符达到一次输出多个数据的目的。例如：

```
name = "xiaowang"
age = 16
print("学生姓名%s,年龄%d" % (name,age))
```

其中，"%""%s""%d"为格式符号，起一次输出多个数据的作用。语句输出结果为"学生姓名 xiaowang 年龄 16"。语句中的引号、圆括号、逗号等都在英文输入法下输入，否则会报错。

Python 中常用的格式符号以及对应的转换类型如表 3-1 所示。

表 3-1　Python 中常用的格式符号及转换类型

格式符号	转 换 类 型
%c	字符
%s	字符串
%d	有符号十进制整数
%u	无符号十进制整数
%o	八进制整数
%x	十六进制整数(小写字母 0x)
%X	十六进制整数(大写字母 0X)
%f	浮点数
%e	科学记数法(小写 'e')
%E	科学记数法(大写 "E")
%g	%f 和 %e 的简写
%G	%f 和 %E 的简写

3) 换行输出

语句中使用"\n"，"\n"后的内容会在另外一行显示。例如：

```
print('123---')
```

"---"会与 123 在一行显示，输出结果为"123---"。

```
print('123\n---')
```

"---"会在 123 的下一行显示，输出结果为

123

---

2. 运算符

1) 算术运算符

Python 在进行混合运算的时候，运算符会根据优先级的顺序进行计算。

优先级顺序：** 优先于 * /，% 优先于 +、-。与生活中的数学运算一样，也可以使用()来处理优先级的问题。

表 3-2 为 Python 中的运算符介绍，以 a = 10，b = 10 为例。

表 3-2 运算符描述及示例

运算符	描述	示 例
+	加	表示两个对象相加，a + b 输出结果 20
-	减	表示负数，或一个数减去另一个数，a - b 输出结果 0
*	乘	表示两个数相乘，或返回一个被重复若干次的字符串，a*b 输出结果 100
/	除	b/a 输出结果 1
//	取整数	返回商的整数部分 1
%	取余数	返回除法的余数 b%a，输出结果 0
**	指数	a**b 为 10 的 10 次方，输出结果 10 000 000 000

2) 赋值运算符

"="为赋值运算符，表示把等号"="右边的结果赋给等号左边的变量，如 number = 3 + 5 * 6 = 33。

3) 复合赋值运算符

多个赋值运算符同时使用构成复合赋值运算符，复合赋值运算符的描述及示例如表 3-3 所示。

表 3-3 复合赋值运算符的描述及示例

复合赋值运算符	描 述	示 例
+=	加法赋值运算符	c+=a 等效于 c=c+a
-=	减法赋值运算符	c-=a 等效于 c=c-a
*=	乘法赋值运算符	c*=a 等效于 c=c*a
/=	除法赋值运算符	c/=a 等效于 c=c/a
%=	取模赋值运算符	c%=a 等效于 c=c%a
**=	幂赋值运算符	c**=a 等效于 c=c**a
//=	取整除赋值运算符	c//=a 等效于 c=c//a

3.2 变量和数据类型

1. 变量的定义

变量是用来存储程序中的数据的，每个变量都有一个独一无二的名字，程序通过读取变量名称就可读取变量所代表的数据。

例如，a = 2，a 为变量名，2 为变量数据。

2. 变量的命名

变量的命名可以是字母、数字、下画线等，字母要严格区分大小写，数字不能位于变量名首位，不能包含空格、@、% 以及 $ 等特殊字符，避免使用以下画线开头的标识符。变量命名要遵守 Python 标识符命名规范，还要避免和 Python 内置函数以及 Python 保留字重名。例如，b_2 = '字符串'，但 2_b = '字符串' 中的变量名 "2_b" 不符合命名规则。

对变量命名时，可以采用驼峰式命名法或下画线命名法。驼峰式命名法又可分为小驼峰式命名法和大驼峰式命名法。

① 小驼峰式命名法(lower camel case)：名称的第一个单词首字母小写，第二个单词首字母大写，如 myWeb、publicProxy。

② 大驼峰式命名法(upper camel case)：名称每个单词的首字母都采用大写方式，如 WebCode、UserName。

下画线命名法是用下画线"_"来连接所有的单词，如 send_buf。

给模块命名时，名称也应尽量短小，并且全部使用小写字母，可以使用下画线命名方法分隔多个字母，如 hard_copy、meta_information 等。给包命名时，名称应尽量短小，全部使用小写字母，不推荐使用下画线，如 com.mr、com.mr.book 等。给类命名时，应采用单词首字母大写的形式。例如定义一个期刊类，可以命名为 Journal。给模块内部的类命名时，可以采用下画线+首字母大写的形式，如 _Book。给函数、类中的属性和方法命名时，应全部使用小写字母，多个单词之间可以用下画线分隔。给常量命名应全部使用大写字母，单词之间可以用下画线分隔。

以下画线开头的标识符在 Python 语言中有特殊含义，例如，以单下画线开头的标识符(如 _width)表示不能直接访问的类属性，其无法通过 from...import* 的方式导入；以双下画线开头的标识符(如 __add)表示类的私有成员；以双下画线作为开头和结尾的标识符(如 __init__)是专用标识符。因此，除非特定需要，应避免使用这类名称。

Python 的内置函数和保留字如表 3-4 和表 3-5 所示，它们不能作为变量名使用。

表 3-4 Python 的内置函数

abs()	delattr()	hash()	memoryview()	set()
all()	dict()	help()	min()	setattr()
any()	dir()	hex()	next()	slicea()
ascii()	divmod()	id()	object()	sorted()
bin()	enumerate()	input()	oct()	staticmethod()
bool()	eval()	int()	open()	str()
breakpoint()	exec()	isinstance()	ord()	sum()
bytearray()	filter()	issubclass()	pow()	super()
bytes()	float()	iter()	print()	tuple()
callable()	format()	len()	property()	type()
chr()	frozenset()	list()	range()	vars()
classmethod()	getattr()	locals()	repr()	zip()
compile()	globals()	map()	reversed()	__import__()
complex()	hasattr()	max()	round()	

表 3-5　Python 的保留字

and	as	assert	break	class	continue
def	del	elif	else	except	finally
for	from	False	global	if	import
in	is	lambda	nonlocal	not	None
or	pass	raise	return	try	True
while	with	yield			

可执行以下命令查看保留字。

import keyword

Print(keyword.kwlist)

输出结果如下：

['False', 'None', 'True', 'and', 'as', 'assert', 'async', 'await', 'break', 'class', 'continue', 'def', 'del', 'elif', 'else', 'except', 'finally', 'for', 'from', 'global', 'if', 'import', 'in', 'is', 'lambda', 'nonlocal', 'not', 'or', 'pass', 'raise', 'return', 'try', 'while', 'with', 'yield']

Python 是严格区分大小写的，保留字也不例外。例如 if 是保留字，但 IF 不是。

3. 变量、数据类型

变量的数据类型按照是否可以改变分为值类型和引用类型。值类型包括数值、字符串和元组；引用类型包括列表、集合和字典。

数据可分为数值和组。数值包括整数、浮点型、布尔值和复数；组包括序列、集合和字典。序列类包括字符串、列表和元组，数据有序，可通过下标索引来读取，可切分；集合值无序、唯一，不能切分；字典由键值对组成，键具有唯一性，不可重复。

（1）Python 变量不需要类型声明，赋值后便可使用。用等号"="来给变量赋值。等号运算符左边是一个变量名，右边是存储在变量中的值。

【例 3-1】

counter = 10 # 赋值整型变量

miles = 100.0 # 浮点型

name = "Smith" # 字符串

print(counter)

print(miles)

print(name)

其中，10，100.0 和"Smith"分别赋值给 counter，miles 和 name 变量。执行例 3-1 代码会输出以下结果：

10

100.0

Smith

（2）多个变量赋值。

Python 允许同时为多个变量赋值。例如，a = b = c = 3 表示创建一个整数型对象，值为 3，三个变量被分配到相同的内存空间。Python 也可以为多个对象指定多个变量。例如，a, b, c = 1, 3, "Eric"表示两个整数型对象 1 和 3 分别分配给变量 a 和 b, 字符串对象"Eric"分配给变量 c。

4. 数据类型转换

常用的数据类型转换函数及描述如表 3-6 所示。

表 3-6 常用的数据类型转换

数据类型转换函数	描 述
int(x[, base])	将 x 转换为一个整数
float(x)	将 x 转换为一个浮点数
Long(x[,base])	将 x 转换为一个长整数
complex(real[imag])	创建一个复数，real 为实部，imag 为虚部
str(x)	将对象 x 转换为字符串
repr(x)	将对象 x 转换为表达式字符串
eval(str)	计算在字符串中的有效 Python 表达式，并返回一个对象
list(s)	将序列 s 转换为一个列表
tuple(s)	将序列 s 转换为一个元组
chr(x)	将一个整数转换为一个 Unicode 字符
ord(x)	将一个字符转换为它的 ASCII 整数值
hex(x)	将一个整数转换为一个十六进制字符串
oct(x)	将一个整数转换为一个八进制字符串
bin(x)	将一个整数转换为一个二进制字符串

使用表 3-6 中的函数可以转换数据类型，如果想给一个元组添加数据，可以先将它转换成列表，然后添加数据，之后把列表转换成元组，转换过程示例如下：

```
a = (5, 6)        # 先将元组 a 转换成列表 b
b = list(a)       # b = [5, 6]
b.append(7)       # b = [5, 6, 7]
a = tuple(b)      # 将添加好数据的列表 b 转换成元组 a, a = (5, 6, 7)
```

3.3 数 值

1. Python 的数值类型

数字数据类型用于存储数值(number)，Python 支持四种不同的数值类型。

(1) 整型(Int): 又称有符号整型，指正整数或负整数。

(2) 长整型(Long Integers/Long): 指无限大小的整数，整数最后一位是一个大写或小写的 L, 可以代表八进制和十六进制。长整型也可以使用小写 i, 但是建议使用大写 L, 避免与数字 1 混淆。长整型类型只存在于 Python2.X 版本中，在 2.2 以后的版本中，整型类数据溢出后会自动转为长整型类型。在Python3.X版本中长整型类型被移除,使用整型替代。

(3) 浮点型(Floating Point Real Values/Float):由整数部分与小数部分组成,浮点型也可以使用科学记数法表示(2.5e2 = 2.5×10^2 = 250)。

(4) 复数(Complex Numbers):复数由实数部分和虚数部分构成,可以用a + bj或者complex(a, b)表示,复数的实部a和虚部b可以是浮点型或整数。

Python中数学运算常用的函数基本都在math模块、cmath模块中。使用math或cmath函数必须先导入。

(1) 查看math包中的内容。命令如下:

import math
print(dir(math))

输出结果如下:

['__doc__', '__loader__', '__name__', '__package__', '__spec__', 'acos', 'acosh', 'asin', 'asinh', 'atan', 'atan2', 'atanh', 'ceil', 'copysign', 'cos', 'cosh', 'degrees', 'e', 'erf', 'erfc', 'exp', 'expm1', 'fabs', 'factorial', 'floor', 'fmod', 'frexp', 'fsum', 'gamma', 'gcd', 'hypot', 'inf', 'isclose', 'isfinite', 'isinf', 'isnan', 'ldexp', 'lgamma', 'log', 'log10', 'log1p', 'log2', 'modf', 'nan', 'pi', 'pow', 'radians', 'remainder', 'sin', 'sinh', 'sqrt', 'tan', 'tanh', 'tau', 'trunc']

(2) 查看cmath包中的内容。命令如下:

import cmath
print(dir(cmath))

输出结果如下:

['__doc__', '__loader__', '__name__', '__package__', '__spec__', 'acos', 'acosh', 'asin', 'asinh', 'atan', 'atanh', 'cos', 'cosh', 'e', 'exp', 'inf', 'infj', 'isclose', 'isfinite', 'isinf', 'isnan', 'log', 'log10', 'nan', 'nanj', 'phase', 'pi', 'polar', 'rect', 'sin', 'sinh', 'sqrt', 'tan', 'tanh', 'tau']

2. Python 常用比较运算符

表 3-7 所示为常用比较运算符及描述,假设变量 a 为 10,变量 b 为 20。

表 3-7 Python 比较运算符

比较运算符	描 述	实 例
==	等于。比较对象是否相等	(a == b)返回 False
!=	不等于。比较两个对象是否不相等	(a != b)返回 True
<>	不等于。比较两个对象是否不相等	(a <> b)返回 True 这个运算符类似 !=
>	大于。返回 x 是否大于 y	(a > b)返回 False
<	小于。返回 x 是否小于 y。所有比较运算符返回 1 表示真,返回 0 表示假。这分别与特殊的变量 True 和 False 等价	(a < b)返回 True
>=	大于等于。返回 x 是否大于等于 y	(a >= b)返回 False
<=	小于等于。返回 x 是否小于等于 y	(a <= b)返回 True

3. Python 常用数学函数

常用数学函数及描述如表 3-8 所示。

表 3-8　Python 常用数学函数

数学函数	描述
abs(x)	返回数字的绝对值，如 abs(-10) 返回 10
ceil(x)	返回数字的上入整数，如 math.ceil(4.1) 返回 5
cmp(x, y)	如果 x < y 返回 -1，如果 x == y 返回 0，如果 x > y 返回 1
exp(x)	返回 e 的 x 次幂(ex)，如 math.exp(1) 返回 2.718 281 828 459 045
fabs(x)	返回数字的绝对值，如 math.fabs(-10) 返回 10.0
floor(x)	返回数字的整数，如 math.floor(4.9) 返回 4
log(x)	如 math.log(math.e) 返回 1.0，math.log(100, 10) 返回 2.0
log10(x)	返回以 10 为基数的 x 的对数，如 math.log10(100) 返回 2.0
max(x1, x2, ...)	返回给定参数的最大值，参数可以为序列
min(x1, x2, ...)	返回给定参数的最小值，参数可以为序列
modf(x)	返回 x 的整数部分与小数部分，两部分的数值符号与 x 相同，整数部分以浮点型表示
pow(x, y)	x**y 运算后的值
round(x [,n])	返回浮点数 x 的四舍五入值，如给出 n 值，则代表舍入到小数点后的位数
sqrt(x)	返回数字 x 的平方根

4. Python 随机数函数

随机数用于数学、游戏、安全等领域，还经常被嵌入到算法中提高算法效率和程序的安全性。Python 常用随机数函数及描述如表 3-9 所示。

表 3-9　Python 常用随机数函数

随机数函数	描述
choice(seq)	从序列的元素中随机挑选一个元素，如 random.choice(range(10))，从 0 到 9 中随机挑选一个整数
randrange ([start,] stop [,step])	从指定范围内，按指定基数递增的集合中获取一个随机数，基数默认值为 1
random()	随机生成下一个实数，在(0, 1)范围内
seed([x])	改变随机数生成器的种子 seed。如果不了解其原理，不必特别去设定 seed，Python 会帮助选择 seed
shuffle(lst)	将序列的所有元素随机排序
uniform(x, y)	随机生成下一个实数，在[x, y]范围内

5. 常用的 Python 三角函数

常用的 Python 三角函数及描述如表 3-10 所示。

表 3-10　常用的 Python 三角函数

三角函数	描　　述
acos(x)	返回 x 的反余弦弧度值
asin(x)	返回 x 的反正弦弧度值
atan(x)	返回 x 的反正切弧度值
atan2(y, x)	返回给定的 x 及 y 坐标值的反正切值
cos(x)	返回 x 的弧度的余弦值
hypot(x, y)	返回欧几里德范数 sqrt(x*x + y*y)
sin(x)	返回的 x 弧度的正弦值
tan(x)	返回 x 弧度的正切值
degrees(x)	将弧度转换为角度，如 degrees(math.pi/2)，返回 90.0
radians(x)	将角度转换为弧度

6. 常见的 Python 数学常量

常见的 Python 数学常量及描述如表 3-11 所示。

表 3-11　常见的 Python 数学常量

数字常量	描　　述
pi	数学常量 pi(圆周率，一般用 π 来表示)
e	数学常量 e，即自然常数

3.4　字　符　串

1. Python 字符串介绍

字符串或串(String)是由数字、字母、下画线等组成的一串字符。

字符串必须使用单引号、双引号或三引号括起来，两边的引号必须对称。通常把字符串放在单引号里，如果字符串本身有单引号，可使用转义符反斜杠"\"加单引号或用双引号把字符串括起来表示；如果字符串本身有双引号，可使用转义符加单引号或用三引号把字符串括起来表示。字符串是不可改变数据，但可以对字符串进行拼接、替换、删除、截取、赋值、连接、比较、查找、分割等操作。

注意：Python 2.X 要求在源程序中增加 "#coding:utf-8" 才能支持中文字符，Python 3.X 对中文字符支持较好。

(1) 用单引号表示字符串的方法。示例如下：

```
str1 = 'Feng Zikai'
str2 = "黄先生是一位有名的画家"
print(str1)
print(str2)
```

(2) 用转义符或双引号表示字符串的方法。若要将 I'm a graduate 作为字符串，不用双引号或转义符 Python 无法识别单引号后面的数据，所以必须使用转义符或双引号。示例如下：

str3 = 'I\'m a graduate' #单引号前加 "\"

str4 = "I'm a graduate"

(3) 用转义符或三引号表示字符串的方法。若字符串本身带有双引号，则使用 "\" 转义符或三引号。示例如下：

str5 = 'The parrot said:\"I can talk and play. Man, why can\'t I teach you?\".'

str6 = '"The parrot said:"I can talk and play. Man, why can't I teach you?".'''

也可以使用以下形式表示字符串：

str6 = '''The parrot said:

"I can talk and play.

Man, why can't I teach you?".'''

Print(str6)

输出结果如下：

The parrot said:

"I can talk and play.

Man, why can't I teach you?".

若字符串本身带有转义符，可采用 "\\" 转义反斜杠 "\"；也可以在字符串引号外加 "r"，这样反斜杠就不会被当成特殊字符。例如：

'F:\\projects\\thesis\\Paper-2\\Dec 16-2020 Revise\\Vader\\Figures 1-11'

或者

r' F:\projects\thesis\Paper-2\Dec 16-2020 Revise\Vader\Figures 1-11'

2. 查看字符串

查看字符串操作命令如下：

print(dir(str))

输出结果如下：

['__add__', '__class__', '__contains__', '__delattr__', '__dir__', '__doc__', '__eq__', '__format__', '__ge__', '__getattribute__', '__getitem__', '__getnewargs__', '__gt__', '__hash__', '__init__', '__init_subclass__', '__iter__', '__le__', '__len__', '__lt__', '__mod__', '__mul__', '__ne__', '__new__', '__reduce__', '__reduce_ex__', '__repr__', '__rmod__', '__rmul__', '__setattr__', '__sizeof__', '__str__', '__subclasshook__', 'capitalize', 'casefold', 'center', 'count', 'encode', 'endswith', 'expandtabs', 'find', 'format', 'format_map', 'index', 'isalnum', 'isalpha', 'isascii', 'isdecimal', 'isdigit', 'isidentifier', 'islower', 'isnumeric', 'isprintable', 'isspace', 'istitle', 'isupper', 'join', 'ljust', 'lower', 'lstrip', 'maketrans', 'partition', 'replace', 'rfind', 'rindex', 'rjust', 'rpartition', 'rsplit', 'rstrip', 'split', 'splitlines', 'startswith', 'strip', 'swapcase', 'title', 'translate', 'upper', 'zfill']

3. 拼接字符串

如果直接将两个字符串紧挨着写在一起，Python 就会自动拼接它们，例如：

s1 = "Hi,"'Python'

print(s1)

输出结果：

Hi, Python

上面这种代码只是书写字符串的一种特殊方法，并不能真正用于拼接字符串。一般可以使用"+"拼接字符串。示例如下：

```
s2 = "Coding "
s3 = "is interesting"
# 使用"+"拼接字符串
s4 = s2 + s3
print(s4)
```

输出结果：Coding is interesting

在实际编程中，有时需要将字符串与数值进行拼接，而 Python 不允许直接拼接数值和字符串，因此必须先使用 str()或 repr()函数将数值转换成字符串再拼接。也可用"str.join"连接2个字符串，可指定连接符号。示例如下：

```
name=["Amy", "Adam"]
s="******".join(name)
print(s)
```

输出结果: Amy******Adam

```
L='python'
L_J = '_'.join(L)
Print(L_J)
```

输出结果: p_y_t_h_o_n

4. 数字和字符串拼接

使用 str()或 repr()函数将数值转换成字符串。示例如下：

```
s1 = "王小民的年龄是: "
p = 99
print(s1 + p)
```

字符串直接拼接数值，程序报错。应先使用 str()或 repr()将数值转换成字符串，示例如下：

```
print(s1 + str(p))
print(s1 + repr(p))
```

str()和 repr()函数都可以将数值转换成字符串，其中 str 是 Python 内置的类型(和 int、float 一样)，repr()则只是一个函数。此外，repr 还会以 Python 表达式的形式来表示值。对比以下代码：

```
st_rp = "I am testing the function repr."
print (st_rp)
print(repr(st_rp))
```

上面代码中的 st_rp 本身就是一个字符串，但程序依然使用了 repr()对字符串进行转换。运行上面程序，输出结果如下：

I am testing the function repr.
'I am testing the function repr.'

由此可以看出，如果直接使用 print()函数输出字符串，只能看到不带引号的字符串内容；但如果先使用 repr()函数对字符串进行处理，再使用 print()执行输出，就可以看到带引号的字符串，这就是字符串的 Python 的表达式形式。

在编辑器中输入一个变量或表达式并执行时，Python 会自动使用 repr()函数处理该变量或表达式。

5. 去除字符串中的空格

去除字符串中空格的命令主要有以下几种。

(1) str.strip()：删除字符串两边的指定字符，括号里写入指定字符，默认为空格。示例如下：

```
a='    python    '
b=a.strip()
print(b)
输出：python
```

(2) str.lstrip()：删除字符串左边的指定字符，括号里写入指定字符，默认空格。示例如下：

```
a='    python    '
b=a.lstrip()
print(b)
输出：python          #python 右边有空格
```

(3) str.rstrip()：删除字符串右边的指定字符，括号里写入指定字符，默认空格。示例如下：

```
a='    python    '
b=a.rstrip()
print(b)
输出：python
```

6. 复制字符串

复制字符串的命令如下：

```
a='hi python'
b=a
print(a,b)
输出：hi python hi python
```

7. 查看字符串长度(len())

查看字符串长度的命令如下：

```
a=' hi python'
print(len(a))
输出：9
```

8. 查看是否包含指定字符串

查看是否包含指定字符串的命令如下：

```
a='hi python'
print('hey' in a)
输出：False
print('snake' not in a)
输出：True
```

9. 转换字符串大小写

转换字符串大小写的命令如下：

```
#S.lower()   #全部转换为小写
a='Hi PYthon'
print(a.lower())
输出：  hi python
# S.upper()   #全部转换为大写
a='hi python'
print(a.upper())
输出：HI PYTHON
# S.swapcase()   #大小写互换
a='hi Python'
print(a.swapcase())
输出：HI pYTHON
# S.capitalize()   #首字母大写
a='hi python'
print(a.capitalize())
输出：Hi python
```

10. 将字符串居中放置在指定宽度的字符串内

调用了字符串对象的 center()方法将字符串居中放置在指定宽度的字符串内，命令如下：

```
# str.center()
a='hi python'
print(a.center(30,'*'))
输出：**********hi python***********
```

字符串 a 被居中放置在一个宽度为 30 的字符串内，并且在两侧用 '*' 填充。由于原字符串的长度为 8，所以左右两侧分别用(30－8)/2 = 11 个 '*' 进行填充。

11. 截取字符串

Python 的字符串列表有 2 种取值顺序。

(1) 从左到右索引默认 0 开始的，最大范围是字符串长度减 1。

(2) 从右到左索引默认 -1 开始的，最大范围是字符串开头。

字符串取值顺序如图 3-1 所示。

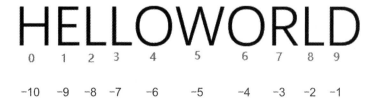

图 3-1　字符串取值顺序图

如果要实现从字符串中截取一段字符串的话，可以使用[头下标:尾下标]的方式截取，下标从 0 开始算起，可以是正数或负数；下标可以为空，表示取到头或尾。[头下标:尾下标]获取的子字符串包含头下标的字符，但不包含尾下标的字符。示例如下：

s = 'abcdefgh'

print(s[2:6])

输出：cdef

结果包含 s[2]的值 c，而取到的最大范围不包括尾下标 s[6]的值 g。更多示例：

st = '0123456789abcde'

print(st[0:5])　#截取第一位到第五位的字符

输出：01234

print(st[:])　#截取字符串的全部字符

输出：0123456789abcde

print(st[5:])　#截取第六个字符到结尾

输出：56789abcde

print(st[:-3])　#截取从头开始到倒数第三个字符之前

输出：0123456789ab

print(st[3])　#截取第四个字符

输出：3

print(st[-1])　#截取倒数第一个字符

输出：e

print(st[::-1])　#创造一个与原字符串顺序相反的字符串

输出：edcba9876543210

print(st[-4:-1])　#截取倒数第四位与倒数第一位之间的字符

输出：bcd

print(st[-3:])　#截取倒数第三位到结尾

输出：cde

print(st[:-6:-3])　#逆序截取，截取倒数第六位数与倒数第三位数之间

输出：eb

12. 切分字符串

(1) 将字符串切分为前中后三部分，示例如下：

s="Amy loves Amy's sisters"

res = s.partition('loves')

print(res)

输出：('Amy ', 'loves', ' Amy's sisters')

(2) 按输入字符切分字符串，示例如下：

s="Amy loves Amy's sisters"

res=s.split("e")

print(res)

输出：['Amy lov', 's Amy's sist', 'rs']

(3) 根据换行符执行切分字符串，示例如下：

```
s="Amy loves\n Amy's \nsisters"
res=s.splitlines()
print(res)
```
输出：['Amy loves', ' Amy's ', 'sisters']

13. 替换字符串

替换字符串的命令：

S.replace(old, new[, count])

将字符串中的 old 替换为 new，如果给定 count，则表示只替换前 count 个 old 字符串。如果在字符串 S 中搜索不到 old，则无法替换，直接返回字符串 S(不创建新字符串对象)。示例如下：

```
s= "Amy loves Amy's sisters"
res = s.replace("Amy", "Dandry")
print(res)
```
输出：Dandry loves Dandry's sisters
```
Res1 = s.replace("Amy", "Dandry", 1)
print(res1)
```
输出：Dandry loves Amy's sisters

14. 查找字符串

查找符串的命令：

S.count(sub[, start[, end]])

返回字符串 S 中子串 sub 出现的次数，可以指定从哪里开始计算(start)以及计算到哪里结束(end)，索引从 0 开始计算，不包括 end 边界。示例如下：

```
S = 'xyabxyxy'
print(S.count('xy'))
# 次数 3，查找范围为所有字符
print(S.count('xy',1))
# 次数 2，从 index=1 算起，即从'y'开始查找，查找的范围为'yabxyxy'
print(S.count('xy',1,7))
# 次数 1，查找从 index=1 开始到 index=7 结束，不包括 7，查找的范围为'yabxyx'
```

3.5 列　　表

1. 列表介绍

列表指的是一种元素以逗号分隔，用中括号标识的可修改的有序序列数据结构类型。列表的格式:变量名=[元素 1，元素 2，元素 3，…]。例如：

```
Lst = [1, 2, 3, 4, 5, 6]
Ls = ['ab', 'c', 'd', 'ef', 'g', 'h']
Lst1 = [2, 4, 6, [1, 2, 3, 4, 5, 6 ]]
```

列表中的元素可以是字符、数字、字符串等，甚至可以是列表，列表中值的切割可以用变量[头下标:尾下标]截取相应的列表元素，从左到右索引默认 0 开始，从右到左索引默认从 −1 开始，下标可以为空，表示截取到头或尾，例如[:8]表示从头截取前 8 个元素，[6:]表示截取第 6 位后的所有列表元素(不含第六位元素)。

列表是序列的一种，列表的序列操作与字符串操作类似，唯一的区别是列表的操作结果是列表而不是字符串。可以对列表进行相加、重复、索引、切片、循环遍历等操作。

2. 查看列表操作命令

查看列表操作命令如下：

print(dir(list))

输出结果：['__add__', '__class__', '__contains__', '__delattr__', '__delitem__', '__dir__', '__doc__', '__eq__', '__format__', '__ge__', '__getattribute__', '__getitem__', '__gt__', '__hash__', '__iadd__', '__imul__', '__init__', '__init_subclass__', '__iter__', '__le__', '__len__', '__lt__', '__mul__', '__ne__', '__new__', '__reduce__', '__reduce_ex__', '__repr__', '__reversed__', '__rmul__', '__setattr__', '__setitem__', '__sizeof__', '__str__', '__subclasshook__', 'append', 'clear', 'copy', 'count', 'extend', 'index', 'insert', 'pop', 'remove', 'reverse', 'sort']

3. 创建列表

(1) list()创建列表。使用 list()可以将任何可迭代的数据转化成列表。示例如下：

a = list()
a = list(range(11))
print(a)
输出： [0, 1, 2, 3, 4, 5, 6, 7, 8, 9, 10]
a = list("python, world")
print(a)
输出：['p', 'y', 't', 'h', 'o', 'n', ',', ' ', 'w', 'o', 'r', 'l', 'd']

(2) range()创建整数列表。range()可以创建整数列表，这在开发中非常有用。语法格式为 range(start，end step)

其中，start 为可选参数，表示起始数字。默认为 0；end 为必选参数，表示结尾数字；step 为可选参数，表示步长，默认为 1。

Python3 中 range()返回的是一个 range 对象，而不是列表。需要通过 list()方法将其转换成列表对象。示例如下：

ls1 = list(range(3,15,3))
print(ls1)
输出：[3, 6, 9, 12]
ls2 = list(range(15,3,-2))
print(ls2)
输出：[15, 13, 11, 9, 7, 5]
ls3 = list(range(3,-10,-3))
print(ls3)
输出：[3, 0, -3, -6, -9]

4. 增加列表元素

给列表增加元素。示例如下：

```
nameList = ['王三','李四','张五','赵二','孙六','Adam','Eric']
nameList.append('孙七')
print('增加第一个元素：',nameList)
输出：增加一个元素：    ['王三', '李四', '张五', '赵二', '孙六', 'Adam', 'Eric', '孙七']
nameList.append('孙八')
print('增加第二个元素：',nameList)
输出：增加第二个元素：    ['王三', '李四', '张五', '赵二', '孙六', 'Adam', 'Eric', '孙七', '孙八']
Lst1 = [1,3, 5, 8, 5, 9, 10, 11, 12, 13]
nameList = ['王三','李四','张五','赵二','孙六','Adam','Eric']
Lst1.append('a1')
print(Lst1)
输出：[1, 3, 5, 8, 5, 9, 10, 11, 12, 13, 'a1']
```

5. insert()插入元素

使用 insert()方法可以将指定的元素插入列表对象的任意制定位置。示例如下：

```
a = [10,20,30]
a.insert(2,100)
print(a)
输出：[10, 20, 100, 30]
```

6. 乘法扩展

使用乘法扩展列表，生成一个新列表。新列表元素是原列表元素的多次重复。示例如下：

```
a = ['xyz',111]
b = a*3
print(b)
输出：['xyz', 111, 'xyz', 111, 'xyz', 111]
```

7. 删除列表元素

(1) Del 方法的下标从 0 开始，因此第二个元素对应列表中的下标 1。示例如下：

```
del nameList[1]
print('删除第二个元素：',nameList)
输出：删除第二个元素：    ['王三', '张五', '赵二', '孙六', 'Adam', 'Eric']
```

(2) remove()方法用于删除首次出现的指定元素，若不存在该元素抛出异常。若存在，只删除第一个找到的元素。示例如下：

```
a = [5,11,10,20,30,40,50,20,30,20,30,40,60]
a.remove(20)
print(a)
输出：[5, 11, 10, 30, 40, 50, 20, 30, 20, 30, 40, 60]
a.remove(80)
```

print(a)

输出：ValueError: list.remove(x): x not in list

8. 查询列表元素

查询列表元素是使用下标索引来访问列表中的值，下标从 0 开始。示例如下：

name1 = nameList[0]

print('查询列表的第一个元素：',name1)

#输出：查询列表的第一个元素： 王三

print(nameList[:5])

输出：['王三', '李四', '张五', '赵二', '孙六']

print(nameList[3:])

输出：['赵二', '孙六', 'Adam', 'Eric']

9. 修改列表元素

修改列表元素要直接指定下标的值修改。示例如下：

print('修改之前的第一个元素是：', nameList[0])

输出：修改之前的第一个元素是： 王三

nameList[0] = '孙大圣'

print('修改之后的第一个元素是：', nameList[0])

输出：修改之后的第一个元素是： 孙大圣

10. 列表排序

(1) 修改原列表，不建新列表的排序。示例如下：

a = [50,20,10,30,40]

print(id(a))

输出：2653159149512

(2) a.sort() # 默认是升序排列。示例如下：

a.sort()

print(a)

输出：[10, 20, 30, 40, 50]

(3) a.sort(reverse=True) # 降序排列。示例如下：

print(a)

输出：[50, 40, 30, 20, 10]

(4) random.shuffle( ) # 打乱顺序。示例如下：

import random

random.shuffle(a)

print(a)

输出：[20, 30, 10, 40, 50]

(5) 建新列表的排序。可以通过内置函数 sorted()进行排序，用这个方法返回新列表，不对原列表做修改。示例如下：

```
a = [70,3,20,10,30,40]
print(id(a))
输出：1846369763784
a = sorted(a) #默认升序
print(a)
输出：[3, 10, 20, 30, 40, 70]
print(id(a))
输出：1846368325576
a = [20,10,30,40]
b = sorted(a)
print(b)
输出：[3, 10, 20, 30, 40, 70]
print(id(b))
输出：1731714007560
c = sorted(a,reverse=True) #降序
print(c)
输出：[70, 40, 30, 20, 10, 3]
print(id(c))
输出：2070053517512
```

通过列表 a, b, c 的"id"值，可以看出生成的列表对象 b 和 c 都是完全新的列表对象。

(6) reversed()# 返回迭代器。内置函数 reversed()支持进行逆序排列，与列表对象 reverse()方法不同的是，内置函数 reversed()不对原列表做任何修改，只返回一个逆序排列的迭代器对象。示例如下：

```
a = [50,10,20,10,30,40]
c = reversed(a)
print(c)
输出： list_reverseiterator object at 0x0000000002BCCEB8>
print(list(c))
输出：[40, 30, 10, 20, 10, 50]
```

11. 列表元素统计

(1) 查看列表长度，示例如下：

```
Lst1 = [1,3, 5, 8, 5, 9, 10, 11, 12, 13]
Lst2 = [3, 6, 6, 8, 9, 11, 14, 12, 9, 3]
Print(len(Lst1))
输出：10
```

(2) 查看列表最值，示例如下：

```
print(max(Lst1))
```

输出：13
print(min(Lst2))
输出：3

为准确查找最大值或最小值，也可以使用以下代码：
import decimal
max_Lst1 = max([decimal.Decimal(i) for i in Lst1])
print(max_Lst1)
输出：13
min_Lst2 = min([decimal.Decimal(i) for i in Lst2])
print(min_Lst2)
输出：3

或者先用"int"把列表里的数值变为整数后再找最大值或最小值，示例如下：
max_value = max([int(i) for i in Lst1])
print(max_value)
输出：13
min_value = min([int(i) for i in Lst1])
print(min_value)
输出：1

如果列表中有 float 类型数据，表达式中使用"float"代替"int"。示例如下：
Lst3 = [1,3, 5, 8, 5, 9, 10, 11, 12, 13, 13.6, 5.8]
max_value = max([float(i) for i in Lst3])
print(max_value)
输出：13.6
count_1 = Lst1.count(3)
# print(count_1)
输出：1
count_2 = nameList.count('赵二')
# print(count_2)
输出：1

(3) 对数值型列表的所有元素进行求和。示例如下。
a = [13,14,20,16,9,19]
sum_a = sum(a)
print(sum_a)
输出：91

12. Python 列表脚本操作符

列表中"+"和"*"操作符与字符串中的操作方法相似。"+"号用于组合列表，"*"号用于重复列表操作。Python 列表脚本操作符如表 3-12 所示。

表 3-12 Python 列表脚本操作符

Python 表达式	结　果	描　述
len([1, 2, 3])	3	长度
[1, 2, 3] + [4, 5, 6]	[1, 2, 3, 4, 5, 6]	组合
['Hi!'] * 4	['Hi!', 'Hi!', 'Hi!', 'Hi!']	重复
3 in [1, 2, 3]	True	元素是否存在列表中
for x in [1, 2, 3]: print x,	1 2 3	迭代

13. Python 列表函数和方法

Python 包含的列表函数及描述如表 3-13 所示。

表 3-13 Python 列表函数

列表函数	描　述
len(list)	列表元素个数
max(list)	返回列表元素最大值
min(list)	返回列表元素最小值
list(seq)	将元组转换为列表

14. Python 列表方法

Python 列表方法及功能描述见表 3-14 所示。

表 3-14 Python 列表方法及功能描述

列表方法	功 能 描 述
list.append(obj)	在列表末尾添加新的对象
list.count(obj)	统计某个元素在列表中出现的次数
list.extend(seq)	在列表末尾一次性追加另一个序列中的多个值(用新列表扩展原来的列表)
list.index(obj)	从列表中找出某个值第一个匹配项的索引位置
list.insert(index, obj)	将对象插入列表
list.pop([index=-1])	移除列表中的一个元素(默认最后一个元素)，并且返回该元素的值
list.remove(obj)	移除列表中某个值的第一个匹配项
list.reverse()	反向选取列表中元素
list.sort(key=None, reverse=False)	对原列表进行排序

3.6　字　典

字典是由键(key)和它对应的值(value)组成的无序数据结构类型，用"{}"标识。

字典中的键和值具有映射关系，字典中的元素通过键来存取，键不允许重复。

可使用花括号语法来创建字典，也可使用 dict()函数来创建字典。使用花括号语法创建字典时，花括号中应包含多个键-值对，键与值之间用英文冒号隔开；多个键-值对之间用英文逗号隔开。

1. 创建字典

(1) 使用花括号语法创建字典，示例如下：

Score_1 = {'音乐': 90, '美术': 89, '数学': 95}

print(score_1)

输出：{'音乐': 90, '美术': 89, '数学': 95}

该字典的键是字符串，值是整数。

(2) 使用元组作为字典的键，示例如下：

dic2 = {(20, 30):'good', 30:'bad'}

print(dic2)

输出：{(20, 30): 'good', 30: 'bad'}

字典中第一个键是元组，第二个键是整数值。

元组可以作为字典的键，但列表不能作为元组的键。这是由于字典要求键必须是不可变类型，列表是可变类型，因此列表不能作为元组的键。

(3) 在使用 dict()函数创建字典时，可以传入多个列表或元组参数作为 key-value 对，每个列表或元组将被当成一个键-值对，因此这些列表或元组都只能包含两个元素。

① 如果不为 dict()函数传入任何参数，则代表创建一个空的字典。创建空的字典的示例代码如下：

dict5 = dict()

print(dict5)

输出：{}

② 创建包含 3 组元组类键-值对的字典。示例如下：

vegetables = [('leek', 1.58), ('cabbage', 1.29), ('lettuce', 2.19)]

dict3 = dict(vegetables)

print(dict3)

输出：{'leek': 1.58, 'cabbage': 1.29, 'lettuce': 2.19}

③ 创建包含 3 组列表类键-值对的字典。示例如下：

cars = [['COVET', 8.5], ['BENZ', 8.3], ['HONDA', 7.9]]

dict4 = dict(cars)

print(dict4)

输出：{'Covet': 8.5, 'BENZ': 8.3, 'HONDA': 7.9} {}

④ 为字典指定关键字参数创建字典。使用关键字参数来创建字典，此时字典的键不允许使用表达式。示例如下：

dict6 = dict(beef = 1.39, pork = 2.59)

print(dict6)

输出：{'beef': 1.39, 'pork': 2.59}

在创建字典时，上面的粗体字代码 key 直接写成 beef、pork，不需要将它们放在引号中。

2. 查询字典操作方法

查询字典操作方法的命令如下：

print(dir(dict))

输出：['__class__', '__contains__', '__delattr__', '__delitem__', '__dir__', '__doc__', '__eq__', '__format__', '__ge__', '__getattribute__', '__getitem__', '__gt__', '__hash__', '__init__', '__init_subclass__', '__iter__', '__le__', '__len__', '__lt__', '__ne__', '__new__', '__reduce__', '__reduce_ex__', '__repr__', '__setattr__', '__setitem__', '__sizeof__', '__str__', '__subclasshook__', 'clear', 'copy', 'fromkeys', 'get', 'items', 'keys', 'pop', 'popitem', 'setdefault', 'update', 'values']

3. 字典的基本用法

需要牢记尽管字典包含多个键-值对，但键是字典的关键数据，因此程序对字典的操作都是基于键的。通过键可以进行增加键-值对、访问值、删除键-值对、更改键-值对、判断指定键-值对是否存在等基本操作。

(1) 增加键-值对：通过给键赋值实现。示例如下：

score_1 = {'音乐': 90, '美术': 89, '数学': 95}

score_1['生物'] = 90

print(score_1)

输出：{'音乐': 90, '美术': 89, '数学': 95, '生物': 90}

原字典中增加了 '生物': 90 键-值对。

(2) 访问值：通过字典中的键访问值。示例如下：

print(score_1['美术'])

输出：89

访问值使用的也是方括号语法，和列表和元组操作类似，只是访问字典值时在方括号中放的是键，而不是列表或元组中的索引。

(3) 删除键-值对：可使用 del 语句删除字典中的键-值对。示例如下：

del score_1['音乐']

del score_1['数学']

print(score_1)

输出：{'美术': 89, '生物': 90}

(4) 更改键-值对：通过字典中的键给现有字典键-值对赋值，可以改变原键-值对；若字典中不存在键，则在字典中新建键-值对。示例如下：

sedans = {'COVET': 20.5, 'BENZ': 8.3, 'HONDA': 7.9}

sedans['BENZ'] = 9.6

'HONDA': 7.9 = 3.8

print(sedans)

也可以使用以下代码同时给字典 sedans 中的 BENZ 和 HONDA 键赋值，并且增加 FORD 键-值对。示例如下：

sedans['BENZ'],sedans['HONDA'], sedans['FORD'] = 9.6,3.8,3.1

print(sedans)

输出：{'COVET': 20.5, 'BENZ': 9.6, 'HONDA': 3.8, 'FORD': 3.1}

(5) 判断指定键值对是否存在：判断字典是否包含指定的键可使用 in 或 not in 运算符。需要指出的是，对于字典而言，in 或 not in 运算符都是基于键来判断的。示例如下：

判断字典 sedans 是否包含名为'HONDA'的键

print('HONDA' in sedans)

输出：True

判断 sedans 是否包含名为'PORSCHE'的键

print('PORSCHE' in sedans)

输出：False

判断 sedans 是否包含名为' LAMBORGHINI'的键

print('LAMBORGHINI' not in sedans)

输出：True

字典的键就相当于列表的索引，不同的是，索引是整数类型，而字典中的键可以是任意不可变类型。字典相当于索引是任意不可变类型的列表，而列表则相当于键只能是整数的字典。因此，如果程序中要使用字典的键都是整数类型，则可考虑换成列表。此外，列表的索引总是从 0 开始连续增大；而字典的键即使是整数类型，也不需要从 0 开始，更不需要连续。另外，列表不允许对不存在的索引赋值，但字典允许直接对原字典中不存在的键赋值。

4. 字典操作的常用方法

(1) fromkeys()方法。fromkeys()方法使用给定的多个键创建字典，这些键对应的值默认都是 None；也可以额外传入一个参数作为默认值。该方法通常会使用"dict"类直接调用。示例如下：

使用列表创建包含 2 个 key 的字典。

a_dict = dict.fromkeys(['a', 'f'])

print(a_dict)

输出：{'a': None, 'f': None}

使用元组创建包含 2 个 key 的字典。

b_dict = dict.fromkeys((2, 15))

print(b_dict)

输出：{2: None, 15: None}

使用元组创建包含 2 个 key 的字典，指定默认的 value。

c_dict = dict.fromkeys((70, 90), 'excellent')

print(c_dict)

输出：c_dict = dict.fromkeys((70, 90), 'excellent')

(2) get()方法。get()方法是根据键来获取值。如果使用该方法访问不存在的键，则会简单地返回 None，不会导致错误。示例如下：

sedans = {'COVET': 20.5, 'BENZ': 8.3, 'HONDA': 7.9}

```
print(sedans.get('COVET'))
```
输出：20.5
```
print(sedans.get('PORSCHE'))
```
输出：None
```
print(sedans['PORSCHE'])    # 注意：使用方括号会报错
```
输出：KeyError

(3) update()方法。update()方法可使用键-值对更新已有字典。在执行 update()方法时，如果原字典中包含对应的键，那么原键对应的值会被新值代替；如果原字典中不包含对应的键-值对，则在字典中添加键-值对。示例如下：

```
sedans = {'COVET': 20.5, 'BENZ': 8.3, 'HONDA': 7.9}
sedans.update({'Ford':4.5, 'PORSCHE': 80.3,'BENZ':10.1})
print(sedans)
```
输出：{'COVET': 20.5, 'BENZ': 10.1, 'HONDA': 7.9, 'Ford': 4.5, 'PORSCHE': 80.3}

示例中原字典已包含'BENZ'键-值对，执行代码更新原值 8.3 为 10.1；原字典中没有"FOR"和"PORSCHE"键-值对，执行代码后添加'Ford': 4.5, 'PORSCHE': 80.3 键-值对。

(4) items()、keys()、values()方法。items()、keys()、values()分别用于获取字典中所有键-值对、所有键和所有值。这三种方法依次返回 dict_items、dict_keys 和 dict_values 对象。为方便使用返回数据，可通过 list()函数把它们转换成列表。在 Python 2.x 中，items()、keys()、values()方法的返回值本来就是列表，可以不用 list()函数进行处理。当然，也可以使用 list()函数处理，处理之后列表还是列表。示例如下：

① 获取字典所有的键-值对，返回一个 dict_items 对象。

```
sedans = {'COVET': 20.5, 'BENZ': 8.3, 'HONDA': 7.9}
itms = sedans.items()
print(type(itms))
```
输出：<class 'dict_items'>
```
print(itms)
```
输出：dict_items([('COVET', 20.5), ('BENZ', 8.3), ('HONDA', 7.9)])

将 dict_items 转换成列表
```
print(list(itms))
```
输出：[('COVET', 20.5), ('BENZ', 8.3), ('HONDA', 7.9)]

访问第 2 个 key-value 对
```
print(list(itms)[1])
```
输出：('BENZ', 8.3)

② 获取字典所有的键，返回一个 dict_keys 对象。

```
kys = sedans.keys()
print(type(kys))
```
输出：<class 'dict_keys'>

将 dict_keys 转换成列表
```
print(list(kys))
```

输出：['COVET', 'BENZ', 'HONDA']
访问第 2 个 key
print(list(kys)[1])
输出：'BENZ'

③ 获取字典所有的值，返回一个 dict_values 对象。
vals = sedans.values()
print(type(vals))
输出：<class 'dict_values'>
print(vals)
输出：dict_values([20.5, 8.3, 7.9])
将 dict_values 转换成列表
print(list(vals))
输出：[20.5, 8.3, 7.9]
访问第 2 个 value
print(list(vals)[1])
输出：8.3

(5) pop( )方法。pop()方法用于获取指定键对应的值，并删除这个键值对。示例如下：
sedans = {'COVET': 20.5, 'BENZ': 8.3, 'HONDA': 7.9}
print(sedans.pop('HONDA'))
输出：7.9
print(sedans)
输出：{'COVET': 20.5, 'BENZ': 8.3}

此程序代码将会获取 "HONDA" 对应的值 7.9，并删除该键-值对。

(6) popitem()方法。popitem()方法用于随机弹出字典中的一个键-值对。正如列表的 pop()方法总是弹出列表中最后一个元素，字典的 popitem()也是弹出字典中最后一个键-值对。由于字典存储键-值对的顺序是不可知的，因此使用者感觉字典的 popitem()方法是随机弹出的，但实际上该方法总是弹出底层存储的最后一个键-值对。示例如下：
sedans = {'COVET': 20.5, 'BENZ': 8.3, 'HONDA': 7.9}
print(sedans)
输出：{'COVET': 20.5, 'BENZ': 8.3, 'HONDA': 7.9}
print(sedans.popitem())
输出：('HONDA', 7.9)
print(sedans)
输出：{'COVET': 20.5, 'BENZ': 8.3}

实际上，由于 popitem 弹出的是一个元组，因此，程序可以通过序列解包的方式用两个变量分别接收键和值。示例如下：
将弹出项的键赋值给 k，值赋值给 v。
sedans = {'COVET': 20.5, 'BENZ': 8.3, 'HONDA': 7.9}
k, v = sedans.popitem()

```
print(k, v)
```
输出：HONDA 7.9

(7) setdefault()方法。setdefault()方法用于根据键来获取对应的键值，当程序要获取的键在字典中不存在时，该方法会先为这个不存在的键设置一个默认的键值，然后再返回该键对应的键值。如果该键-值存在，则直接返回该键对应的值。总之，setdefault()方法总会返回指定键对应的键值。示例如下：

```
sedans = {'COVET': 20.5, 'BENZ': 8.3, 'HONDA': 7.9}
print(sedans.setdefault('PORSCHE', 10.2))
```
输出：10.2
```
print(sedans)
```
输出：{'COVET': 20.5, 'BENZ': 8.3, 'HONDA': 7.9, 'PORSCHE': 10.2}
设置默认值，该键在字典中存在，不会修改字典中该键所对应的值
```
print(sedans.setdefault('HONDA', 3.4))
```
输出：7.9
```
print(sedans)
```
输出：{'COVET': 20.5, 'BENZ': 8.3, 'HONDA': 7.9, 'PORSCHE': 10.2}

(8) clear()方法。clear()用于清空字典中所有的键-值对，使用该方法后，该字典就会被清空。示例如下：

```
sedans = {'COVET': 20.5, 'BENZ': 8.3, 'HONDA': 7.9}
print(sedans)
```
输出：{'BMW': 8.5, 'BENZ': 8.3, 'AUDI': 7.9}
清空 sedans 所有键-值对
```
sedans.clear()
print(sedans)
```
输出：{}

5. 使用字典格式化字符串

使用字典格式化字符串是在每个转换说明符中的"%"字符后加上用圆括号括起来的键，再跟上其他说明元素，示例如下：

```
students={'王五':'1001','章辉':'1002','强生':'1003','李刚':'1004'}
print('强生的学号是：%(强生)s'%students)
```
输出：强生的学号是：1003

在格式化字符串时，如果要格式化的字符串模板中包含多个变量，就需要按顺序给出这些变量，当字符串模板中变量数量不大时，该方法是合适的，但如果字符串模板中变量很多，这种按顺序提供变量的方式则有些不合适。可改为在字符串模板中按键指定变量，然后通过字典为字符串模板中的键设置值。示例如下：

```
temp = '商品名称:%(name)s, 价格:%(price)010.2f, 生产厂商:%(manufacturer)s'
book = {'name':'天空之水', 'price': 1000, 'manufacturer': '天地人公司'}
print(temp % book)
```
输出：商品名称:天空之水, 价格:0001000.00, 生产厂商:天地人公司

3.7 集　　合

集合是由不同元素组成的无序、不可变的数据结构类型。通过相应的操作符或方法，集合之间可进行数学集合运算。集合元素可以是字符、字符串或数字等，由花括号来界定。示例如下：

S = {'a','book',1, 2, 5, 6}
print(S)
输出：{1, 2, 5, 6, 'book', 'a'}

1. 集合内置函数

查看集合内置函数的命令如下：

print(dir(set))
输出：['__and__', '__class__', '__contains__', '__delattr__', '__dir__', '__doc__', '__eq__', '__format__', '__ge__', '__getattribute__', '__gt__', '__hash__', '__iand__', '__init__', '__init_subclass__', '__ior__', '__isub__', '__iter__', '__ixor__', '__le__', '__len__', '__lt__', '__ne__', '__new__', '__or__', '__rand__', '__reduce__', '__reduce_ex__', '__repr__', '__ror__', '__rsub__', '__rxor__', '__setattr__', '__sizeof__', '__str__', '__sub__', '__subclasshook__', '__xor__', 'add', 'clear', 'copy', 'difference', 'difference_update', 'discard', 'intersection', 'intersection_update', 'isdisjoint', 'issubset', 'issuperset', 'pop', 'remove', 'symmetric_difference', 'symmetric_difference_update', 'union', 'update']

2. 集合运算

集合元素可以是列表，但不能重复，集合间可以进行并集、交集、子集、超集和差集等运算。示例如下：

set1 = {1,2,3,5,6,7,1,6,3,8,12}
print(set1)
输出：{1, 2, 3, 5, 6, 7, 8, 12}
set2 = {2,3,4,5,12,14,16}
print(set2)
输出：{2,3,4,5,12,14,16}
set3 = set([2,3,5,7,9,5,12,15])
print(set3)
输出：{2, 3, 5, 7, 9, 12, 15}

(1) 并集（"|"）或方法 union()。示例如下：

print(set1|set2)
输出：{1, 2, 3, 4, 5, 6, 7, 8, 12, 14, 16}
print(set1 | set3)
输出：{1, 2, 3, 5, 6, 7, 8, 9, 12, 15}
或

print(set1.union(set2))

输出：{1, 2, 3, 4, 5, 6, 7, 8, 12, 14, 16}

(2) 交集（"&"）或方法 intersection()。示例如下：

print(set1 & set2)

输出：{2, 3, 12, 5}

print(set1.intersection(set2))

输出：{2, 3, 12, 5}

(3) 集合元素为字符串时并集、交集运算。示例如下：

setA = {'about','approx','bat','wang','oh','rock','pad'}

setB = {'roc','and','grain','bat','oh','of','rice'}

print(setA & setB)

输出：{'bat', 'oh'}

print(setA | setB)

输出：{'approx', 'and', 'bat', 'roc', 'rice', 'about', 'of', 'pad', 'rock', 'oh', 'grain', 'wang'}

print(setA.intersection(setB))

输出：{'bat', 'oh'}

(4) 判断子集（"<"）或方法 issubset()。示例如下：

set4 = set([2,6,7,9,12,14,16,15,17,6,8])

print(set4)

输出：{2, 6, 7, 8, 9, 12, 14, 15, 16, 17}

set5 = set([3,5,6,8,12,14,16,18,19,20,22])

print(set5)

输出：{3, 5, 6, 8, 12, 14, 16, 18, 19, 20, 22}

print(set4<set5)

输出：False

print(set4.issubset(set5))

输出：False

print(set5.issubset(set4))

输出：False

(5) 判断超集（">"）或方法 issuperset( )。示例如下：

print(set5>set4)

输出：False

print(set5.issuperset(set4))

输出：False

(6) 差集（'-'）或方法 difference()。示例如下：

Print(set5-set4)

输出：{3, 5, 18, 19, 20, 22}

print(set5.difference(set4))

输出：{3, 5, 18, 19, 20, 22}

(7) 对称差("^")方法或 symmetric_difference()。

两个集合的对称差是只属于其中一个集合,而不属于另一个集合的元素组成的集合。判断对称差时使用"^"操作符,也可使用方法 symmetric_difference()完成。示例如下:

```
print(set5 ^ set4)
# {2, 3, 5, 7, 9, 15, 17, 18, 19, 20, 22}
print(set5.symmetric_difference(set4))
# {2, 3, 5, 7, 9, 15, 17, 18, 19, 20, 22}
```

(8) 判断是否相等。示例如下:

```
print(set2 == set4)
```
输出:False

3. 集合操作的常用方法示例代码

(1) add( )方法向集合中添加元素。示例如下:

```
s = {1, 2, 3, 4, 5, 6,7}
s.add("a")    #向集合中添加元素
print(s)
```
输出:{1, 2, 3, 4, 5, 6, 7, 'a'}

(2) clear( )方法。示例如下:

```
s = {1, 2, 3, 4, 5, 6,7}
s.clear()     #清空集合
print(s)
```
输出:set()

(3) copy( )方法。示例如下:

```
s = {1, 2, 3, 4, 5, 6,7}
new_s = s.copy()    #复制原集合
print(new_s)
```
输出:{1, 2, 3, 4, 5, 6, 7}

(4) remove( )方法。删除集合中的一个元素(如果元素不存在,会返回 KeyError)。示例如下:

```
s = {1, 2, 3, 4, 5, 6,7}
s.remove(3)
print(s)
```
输出:{1, 2, 4, 5, 6, 7}

(5) discard( )方法。删除集合中的一个元素(如果元素不存在,则输出原集合)。示例如下:

```
s = {1, 2, 3, 4, 5, 6,7}
s.discard(5)
print(s)
```
输出:{1, 2, 3, 4, 6, 7}

```
s.discard("abbby")
print(s)
```

输出：{1, 2, 3, 4, 5, 6, 7}

(6) update( )方法。用原集合和另一个集合的并集来更新原集合。例如：

s = {'o', 'p'}

① 添加多个元素更新集合。示例如下：

s.update(['p', 'h', 'i', 'y'])

print(s)

输出：{'i', 'o', 'y', 'p', 'h'}

② 添加列表和集合更新集合。示例如下：

s.update(['Z', 'O'], {'y', 'Y', 'S'})

print(s)

输出：{'y', 'O', 'i', 'Z', 'S', 'h', 'Y', 'p', 'o'}

(7) intersection_update( )方法。用原集合与另一个的交集来更新原集合。示例如下：

s = {'ccd', 't', 'y', 'qd', 'q'}

s2 = {'t', 'h', 'b', 's'}

s.intersection_update(s2)   # 相当于 s = s - s2

print(s)

输出：{'t'}

(8) isdisjoint( )方法。如果两个集合有一个空交集，返回 True；若是非空交集，返回 False。示例如下：

s = {3, 5, 7, 'aa'}

s1 = {6, 9, 11, 'bb'}

s2 = {7, 3, 8, 'cc'}

print(s.isdisjoint(s1))

输出：True

print(s.isdisjoint(s2))

输出：False

(9) difference_update( )方法。从当前集合中删除另一个集合的所有元素。示例如下：

s1 = {6, 7, 1, 2, 3}

s2 = {9, 1, 2, 21, 3, 4, 11}

s3 = {32, 11}

s2.difference_update(s1)   #从 s2 集合中删除 s1 集合中的所有元素

print(s2)

输出：{4, 9, 11, 21}

(10) symmetric_difference_update( )方法。示例如下：

s1 = {6, 7, 1, 2, 3}

s2 = {9, 1, 2, 21, 3, 4, 11}

s1.symmetric_difference_update(s2)   #用 s1 与 s2 两集合的对称差更新 s1 集合

print(s1)

输出：{4, 6, 7, 9, 11, 21}

4. 集合与内置函数

常见集合内置函数及其描述如表 3-15 所示。

表 3-15　集合内置函数及描述

集合内置函数	描述
all()	如果集合中的所有元素都是 True（或集合为空），则返回 True
any()	如果集合中的所有元素都是 True，则返回 True；如果集合为空，则返回 False
enumerate()	返回一个枚举对象，其中包含集合中所有元素的索引和对应的值
len()	返回集合的长度
max()	返回集合中的最大项
min()	返回集合中的最小项
sorted()	从集合中的元素返回新的排序列表，原集合不变
sum()	返回集合的所有元素之和

3.8　元　　组

元组是用小括号界定的不可改变的一种序列，与列表类似，元组中可储存字符串、数字、列表等，各元素间用逗号分开。

1. 元组的形式及示例代码

（1）含数字的元组。元组中只包含一个元素时，需要在元素后面添加逗号来消除歧义。示例如下：

tup0 = (30,)

print(tup0)

输出：30

tup = (2,5,6,9,10)

print(tup)

输出：(2, 5, 6, 9, 10)

（2）含数字和字符串的元组。示例如下：

tup1 = (3, 4, 5, 8, 'we','our','his')

print(tup1)

输出：(3, 4, 5, 8, 'we', 'our', 'his')

（3）含数字、字符串和布尔值 False 的元组。示例如下：

tup2 = (3, 4, 5, 8, 'we','our','his',False)

print(tup2)

输出：(3, 4, 5, 8, 'we', 'our', 'his', False)

（4）含元组的元组。示例如下：

tup01 = (4,6,7,8,(2,3))

```
print(tup01)
```
输出：(4, 6, 7, 8, (2, 3))
tup02 = ((4,6,7,8),(2,3))
```
print(tup02)
```
输出：((4, 6, 7, 8), (2, 3))

(5) 含列表的元组。示例如下：

tup3= ([2,3,5])
```
print(tup3)
```
输出：[2, 3, 5]

(6) 含元组与列表的元组。示例如下：

tup4 = ((4,6,7,8),[2,3])
```
print(tup4)
```
输出：((4, 6, 7, 8), [2, 3])

(7) 句子形式的元组。示例如下：

tup5 = ('Teddy is faithful dog')
```
Print(tup5)
```
输出：Teddy is faithful dog
```
print(tuple(tup5))
```
输出：('T', 'e', 'd', 'd', 'y', ' ', 'i', 's', ' ', 'f', 'a', 'i', 't', 'h', 'f', 'u', 'l', ' ', 'd', 'o', 'g')

2. 元组操作及示例代码

(1) 查看元组。示例如下：

```
print(dir(tuple)
```
输出：['__add__', '__class__', '__contains__', '__delattr__', '__dir__', '__doc__', '__eq__', '__format__', '__ge__', '__getattribute__', '__getitem__', '__getnewargs__', '__gt__', '__hash__', '__init__', '__init_subclass__', '__iter__', '__le__', '__len__', '__lt__', '__mul__', '__ne__', '__new__', '__reduce__', '__reduce_ex__', '__repr__', '__rmul__', '__setattr__', '__sizeof__', '__str__', '__subclasshook__', 'count', 'index']

(2) 元组合并。元组中的元素值是不能修改的，但可以对元组进行连接组合。示例如下：

tup_1=(22,11.35)
tup_2=('aabbc','xxxyz','zzzf')
tup_3=tup_1+tup_2
```
print(tup_3)
```
输出：(22, 11.35, 'aabbc', 'xxxyz', 'zzzf')

(3) 元组中添加元素。示例如下：

tup_4 = ('张','无','忌')
tup_5 = tup_4[0:2]+('三峰',)+ tup_4[2:]
```
print(tup_5)
```
输出：('张', '无', '三峰', '忌')

(4) 删除整个元组。示例如下：

```
tuple1 = (1,2,3,4)
del tuple1
print(tuple1)
输出：NameError: name 'tuple1' is not defined
```

(5) 元组的复制。示例如下：

```
tup5 = ('Teddy is faithful dog')
print(tup5*2)
输出：Teddy is faithful dogTeddy is faithful dog
print(tuple(tup5*2))
输出：('T', 'e', 'd', 'd', 'y', ' ', 'i', 's', ' ', 'f', 'a', 'i', 't', 'h', 'f', 'u', 'l', ' ', 'd', 'o', 'g', 'T', 'e', 'd', 'd', 'y', ' ', 'i', 's', ' ', 'f', 'a', 'i', 't', 'h', 'f', 'u', 'l', ' ', 'd', 'o', 'g')
```

(6) 判断元组中元素是否存在。示例如下：

```
tuple = (1,2,3,4,5,6)
print(1 in tuple)
输出：True
print(10 in tuple)
输出：False
tuple = (1,2,3,4,6,7)
print(1 not in tuple)
输出：False
print(5 not in tuple)
输出：True
```

(7) 元组反转。

① 反转一个元组时，可以使用切片操作符，将步长设置为 -1。Python 将从元组最后一个元素开始向前取元素，直到元组的第一个元素为止，这样就实现了元组的反转。示例如下：

```
tp = (1,2,3,45,7,9)
print(tp[::-1])
输出：(9, 7, 45, 3, 2, 1)
```

② 使用 reversed 内置函数反转元组。示例如下：

```
x = (1, 2, 3, 4)
x = tuple(reversed(x))
print(x)
输出：(4, 3, 2, 1)
```

如果只想遍历元组，可以直接使用 reversed 返回的迭代器，无需再次将其转换为元组。例如：

```
for k in reversed(x):
    print(k)
```

(8) 元组索引、截取。元组也是一个序列，不仅可以访问元组中指定位置的元素，也可以截取索引中的一段元素，示例如下：

```
tup6=('王','张','杨','李','Adam','Smith')
print(tup6[-2])
输出：Adam
print(tup6[:])
输出：('王', '张', '杨', '李', 'Adam', 'Smith')
print(tup6[:-1])
输出：('王', '张', '杨', '李', 'Adam')
```

(9) 元组拆包。示例如下：

```
tup7 = ('ming')
a,b,c,d = tup7
print(a,b)
输出：m i
print(c)
输出：n
num = (1,2,3)
a,b,c = num
print(a,b,c)
输出：1 2 3
```

(10) 元组计数。示例如下：

```
tup8 = (4, 6, 7, 8, 2, 3,6,8,3,2)
print(tup8.count(8))
输出：2
tup9 = ("哈",'哈','哈哈','哦','哦哦','啊啊','嗯','嗯嗯','啊哈')
print(tup9.count("哈"))
输出：2
print(tup9.count('嗯'))
输出：1
```

(11) 元组内置函数。

① len(tuple)：计算元组元素个数。

② max(tuple)：返回元组中元素最大值。

③ min(tuple)：返回元组中元素最小值。

示例如下：

```
print(len(tup9))
输出：9
print(max(tup8))
输出：8
print(min(tup8))
输出：2
```

3.9 推 导 式

Python 推导式是一种独特的数据处理方式,它是可以从一个数据序列构建另一个新的数据序列的结构体。Python 支持各种数据结构的推导式,包括列表推导式、字典推导式、集合推导式和元组推导式。

推导式通常是由 for, in, if 组合的语句,列表推导式、字典推导式、集合推导式和元组推导式基本结构如下:

(1) 列表推导式:[表达式 for 变量 in 列表]或者[表达式 for 变量 in 列表 if 条件]。
(2) 字典推导式:{键语句:值语句 for 值 in 集合}或{键语句:值语句 for 值 in 集合 if 条件}。
(3) 集合推导式:{语句 for 项 in 可迭代的对象}或{语句 for 项 in 可迭代的对象 if 条件}。
(4) 元组推导式:(语句 for 项 in 可迭代的对象)或(语句 for 项 in 可迭代的对象 if 条件)。
可迭代的对象指列表、元组、字典或集合。

1. if 语句

(1) if 用来判断,格式为

if 判断条件:

条件成立,则执行语句。示例如下:

gender = "male"

if gender is "male":

　　print("男性")

输出:男性

(2) 当判断条件为多个值时,可以使用 elif 表示多个条件。示例如下:

score = 75

if score >= 90:

　　print("优秀")

elif score >= 80:

　　print("良好")

elif score >= 70:

　　print("中等")

elif score >= 60:

　　print("及格")

输出:中等

(3) if 常与 else 连用,else 为可选语句。当 if 判断条件不成立时,则执行 else 之后的语句。示例如下:

gender = "female"

if gender is "male":

　　print("男性")

else:

```
    print("待确定")
```
输出：待确定

(4) if、elif 和 else 同时使用。示例如下：

```
age = 151
if age < 10:
    print("儿童")
elif age >= 10 and age<18:
    print("大儿童")
elif age >= 18 and age<150:
    print("成年")
else:
    print("神仙")
```
输出：神仙

2. For 语句

For 语句用于遍历字符串、列表、元组和字典里的元素。要注意 for 语句后有冒号，下一行语句需缩进。

(1) 遍历字符串。示例如下：

```
for char in "python":
    print(char)
```
输出：
p
y
t
h
o
n

(2) 遍历列表元素。示例如下：

```
lst = [1, 6, 9, 8, 'abcd']
for i in lst:
    print(i)
```
输出：
1
6
9
8
abcd

(3) 遍历元组元素。示例如下：

```
tpl = (2, 4, 8, 3, 9)
for i in tpl:
```

```
        print(i)
```
输出：
2
4
8
3
9

(4) 遍历字典键和值。示例如下：
```
dict1 = {'1':'小强','2':'阿强','3':'强生'}
for k in dict1:
        print(k)
```
输出：
1
2
3
```
        print(dict1[k])
```
输出：
小强
阿强
强生

Print 时使用 end=" "，表示遍历字符串后在同一行显示遍历内容，例如：
```
s = "python in the world"
for i in s:
        print(i,end=" ")
```
输出：python in the world

3. 推导式

推导式(Comprehensions)也称解析式，是一种简洁的语法结构，用于快速创建数据结构，如列表、字典和集合等，是 Python 的一种独有特性。它们提供了一种在单个可读性高的语句中构建数据结构的方法。

推导式语法： [返回值 for 元素 in 可迭代对象 if 条件]

其中，推导式使用中括号[]，内部是 for 循环，if 条件语句可选，返回一个新的列表。

常见的推导式类型有以下几种。

(1) 列表推导式。示例如下：
```
range1 = range(10)
list1 = [x * 3 for x in range1]
print(list1)
```
输出：[0, 3, 6, 9, 12, 15, 18, 21, 24, 27]

推导式的作用和结果与以下语句基本相同，不同点在于推导式输出的是列表类型数据，而以下语句输出的是整数类型数据。

```
range1 = range(10)
for x in range1:
    print(x*3,end=' ')
```
输出：0 3 6 9 12 15 18 21 24 27

(2) 嵌套生成列表推导式。示例如下：

```
e_lst = [[x, y, z] for x in range(3) for y in range(5) for z in range(4)]
# e_list 列表包含 60 个元素
print(len(e_lst))
# 60
print(e_lst)
```
输出：[[0, 0, 0], [0, 0, 1], [0, 0, 2], [0, 0, 3], [0, 1, 0], [0, 1, 1], [0, 1, 2], [0, 1, 3], [0, 2, 0], [0, 2, 1], [0, 2, 2], [0, 2, 3], [0, 3, 0], [0, 3, 1], [0, 3, 2], [0, 3, 3], [0, 4, 0], [0, 4, 1], [0, 4, 2], [0, 4, 3], [1, 0, 0], [1, 0, 1], [1, 0, 2], [1, 0, 3], [1, 1, 0], [1, 1, 1], [1, 1, 2], [1, 1, 3], [1, 2, 0], [1, 2, 1], [1, 2, 2], [1, 2, 3], [1, 3, 0], [1, 3, 1], [1, 3, 2], [1, 3, 3], [1, 4, 0], [1, 4, 1], [1, 4, 2], [1, 4, 3], [2, 0, 0], [2, 0, 1], [2, 0, 2], [2, 0, 3], [2, 1, 0], [2, 1, 1], [2, 1, 2], [2, 1, 3], [2, 2, 0], [2, 2, 1], [2, 2, 2], [2, 2, 3], [2, 3, 0], [2, 3, 1], [2, 3, 2], [2, 3, 3], [2, 4, 0], [2, 4, 1], [2, 4, 2], [2, 4, 3]]

(3) 字典推导式。示例如下：

```
old_dic = {'a': 8, 'b': 20,'c':15,'d':9}
new_dic = {k: v for k, v in old_dic.items() if v>10}
print(new_dic)
```
输出：{'b': 20, 'c': 15}

以上推导式转换为 for、if 语句，示例如下：

```
for k, v in old_dic.items():
    if v > 10:
        print(k,v)
```
输出：b 20
　　　c 15

(4) 集合推导式。示例如下：

```
demo_set = {3, 6, 8, 12, 11}
```
遍历以上集合，如果集合元素值可以被 2 整除，则输出元素的 3 次方，推导式如下：

```
new_set = {x**3 for x in demo_set if x %2==0}
print(new_set)
```
输出：{216, 512, 1728}

使用 if、for 语句表示如下：

```
for x in demo_set:
    if x%2==0:
        print(x**3,end = ' ')
```
输出：216 512 1728

(5) 生成器表达式。生成器表达式是一种特殊的推导式，用于创建生成器对象。生成器表达式不会一次性生成所有元素，而是在需要时逐个生成，从而节省内存。

生成器表达式语法：(返回值 for 元素 in 可迭代对象 if 条件)。它们与列表推导式的语法类似，列表解析式的中括号换成小括号，返回一个生成器，示例如下：

a = (x for x in range(1,10))

print(a)

输出：\<generator object \<genexpr> at 0x0000027F9696EEC8>

for x in a:

    print(x,end = ', ')

输出：1, 2, 3, 4, 5, 6, 7, 8, 9,

print(tuple(a))

输出：(1, 2, 3, 4, 5, 6, 7, 8, 9)

3.10 函　　数

函数是为实现单一或相关功能而组织好的可重复利用的代码段，Python 有数学运算、集合类、逻辑判断、输入输出操作等内置函数，用户也可以自定义函数。

1. Python 内置函数

Python 内置函数如表 3-16 所示。

表 3-16　Python 内置函数

abs()	delattr()	hash()	memoryview()	set()
all()	dict()	help()	min()	setattr()
any()	dir()	hex()	next()	slice()
ascii()	divmod()	id()	object()	sorted()
bin()	enumerate()	input()	oct()	staticmethod()
bool()	eval()	int()	open()	str()
breakpoint()	exec()	isinstance()	ord()	sum()
bytearray()	filter()	issubclass()	pow()	super()
bytes()	float()	iter()	print()	tuple()
callable()	format()	len()	property()	type()
chr()	frozenset()	list()	range()	vars()
classmethod()	getattr()	locals()	repr()	zip()
compile()	globals()	map()	reversed()	__import__()
complex()	hasattr()	max()	round()	

2. 数学运算类函数

Python 数学运算类函数及描述如表 3-17 所示。

表 3-17　数学运算类函数及描述

数学运算类函数	描　　述
abs(x)	求绝对值。参数可以是整型，也可以是复数；若参数是复数，则返回复数的模
complex([real[, imag]])	创建一个复数
divmod(a, b)	分别取商和余数，整型、浮点型都可以
float([x])	将一个字符串或数转换为浮点数。如果无参数将返回 0.0
int([x[, base]])	将一个字符转换为 int 类型，base 表示进制
long([x[, base]])	将一个字符转换为 long 类型
pow(x, y[, z])	返回 x 的 y 次幂
range([start], stop[, step])	产生一个序列，默认从 0 开始
round(x[, n])	四舍五入
sum(iterable[, start])	对集合求和
oct(x)	将一个数字转化为八进制
hex(x)	将整数 x 转换为十六进制字符串
chr(i)	返回整数 i 对应的 ASCII 字符
bin(x)	将整数 x 转换为二进制字符串
bool([x])	将 x 转换为 Boolean 类型

3. 集合类操作函数

Python 集合类操作函数及描述如表 3-18 所示。

表 3-18　集合类操作函数及描述

集合类函数	描　　述
basestring()	str 和 unicode 的超类，不能直接调用，可以用作 isinstance 判断
format(value [, format_spec])	格式化输出字符串,格式化的参数顺序从 0 开始,如"I am {0}, I like {1}"
unichr(i)	返回给定 int 类型的 unicode
enumerate(sequence [, start = 0])	返回一个可枚举的对象,该对象的 next()方法将返回一个 tuple
iter(o[, sentinel])	生成一个对象的迭代器，第二个参数表示分隔符
max(iterable[, args...][key])	返回集合中的最大值

集合类函数	描 述
min(iterable[, args...][key])	返回集合中的最小值
dict([arg])	创建数据字典
list([iterable])	将一个集合类转换为另外一个集合类
set()	set 对象实例化
frozenset([iterable])	产生一个不可变的 set
str([object])	转换为 string 类型
sorted(iterable[, cmp[, key[, reverse]]])	对集合排序
tuple([iterable])	生成一个 tuple 类型
xrange([start], stop[, step])	xrange()函数与 range()类似,但 xrnage()并不创建列表,而是返回一个 xrange 对象,它的行为与列表相似,但是只在需要时才计算列表值,当列表很大时,这个特性能节省内存

4. 逻辑判断类函数

Python 逻辑判断类函数及描述如表 3-19 所示。

表 3-19 逻辑判断类函数

逻辑判断类函数	描 述
all(iterable)	集合中的元素都为真的时候为真,若为空串则返回为 True
any(iterable)	集合中的元素有一个为真的时候为真,若为空串则返回为 False
cmp(x, y)	如果 x < y,返回负数;若 x == y,返回 0;若 x > y,返回正数

5. 反射函数

Python 反射函数及描述如表 3-20 所示。

表 3-20 反 射 函 数

反射函数	描 述
callable(object)	检查对象 object 是否可调用 (1) 类是可以被调用的。 (2) 实例是不可以被调用的,除非类中声明了 __call__ 方法
classmethod()	(1) 注解,用来说明这个方式是个类方法。 (2) 类方法,既可以被类调用,也可以被实例调用。 (3) 类方法,类似于 Java 中的 static 方法。 (4) 类方法中不需要有 self 参数
compile(source, filename)	将 source 编译为代码或者 AST 对象。代码对象能够通过 exec 语句来执行,或者 eval()进行求值

续表一

反射函数	描 述
mode[, flags[, dont_inherit]])	(1) 参数 source，字符串或者 AST(Abstract Syntax Trees)对象。 (2) 参数 filename，代码文件名称，如果不是从文件读取代码则传递一些可辨认的值。 (3) 参数 model，指定编译代码的种类。可以指定为'exec''eval''single'。 (4) 参数 flag 和 dont_inherit，用来控制编译源码
dir([object])	(1) 不带参数时，返回当前范围内的变量、方法和定义的类型列表。 (2) 带参数时，返回参数的属性、方法列表。 (3) 当参数为实例时，如果参数包含方法 __dir__()，该方法将被调用。 (4) 如果参数不包含 __dir__()，该方法将最大限度地收集参数信息
delattr(object, name)	删除 object 对象名为 name 的属性
eval(expression [, globals [, locals]])	计算表达式 expression 的值
execfile(filename [, globals [, locals]])	用法类似 exec()，不同的是 execfile 的参数 filename 为文件名，而 exec 的参数为字符串
filter(function, iterable)	构造一个序列，等价于[item for item in iterable if function(item)] (1) 参数 function，返回值为 True 或 False 的函数，可以为 None。 (2) 参数 iterable，序列或可迭代对象
getattr(object, name [, defalut])	获取一个类的属性
globals()	返回一个描述当前全局符号表的字典
hasattr(object, name)	判断对象 object 是否包含名为 name 的特性
hash(object)	如果对象 object 为哈希表类型，返回对象 object 的哈希值
id(object)	返回对象的唯一标识
isinstance(object, classinfo)	判断 object 是否是 class 的实例
issubclass(class, classinfo)	判断是否为子类
len(s)	返回集合长度
locals()	返回当前的变量列表
map(function, iterable, ...)	遍历每个元素，执行 function 操作
memoryview(obj)	返回一个内存镜像类型的对象
next(iterator[, default])	类似 iterator.next()
object()	基类

续表二

反射函数	描 述
property([fget[, fset[, fdel[, doc]]]])	属性访问的包装类，设置后可以通过 c.x=value 等来访问 setter 和 getter
reduce(function, iterable[, initializer])	调用可选的初始参数执行函数操作；如果没有初始值，那么从可迭代对象的第一个元素开始进行函数的归约操作，最终得到一个单一的输出
reload(module)	重新加载模块
setattr(object, name, value)	设置属性值
repr(object)	将一个对象变换为可打印的格式
slice()	实现切片
staticmethod	声明静态方法，是个注解
super(type[, object-or-type])	引用父类
type(object)	返回该 object 的类型
vars([object])	返回对象的变量，若无参数与 dict()方法类似
bytearray([source [, encoding [, errors]]])	返回一个 byte 数组 (1) 如果 source 为整数，则返回一个长度为 source 的初始化数组。 (2) 如果 source 为字符串，则按照指定的 encoding 将字符串转换为字节序列。 (3) 如果 source 为可迭代类型，则元素必须为[0, 255]中的整数。 (4) 如果 source 为与 buffer 接口一致的对象，则此对象也可以被用于初始化 bytearray
zip([iterable, ...])	用于将可迭代的对象作为参数，将对象中对应的元素打包成一个个元组，然后返回由这些元组组成的列表

6. 输入输出操作类函数

Python 输入输出类函数及描述如表 3-21 所示。

表 3-21 输入输出类函数及描述

输入输出类函数	描 述
file(filename [, mode [, bufsize]])	file 类型构造函数的作用是为打开一个文件，如果文件不存在且 mode 为写或追加时，文件将被创建。添加 'b' 到 mode 参数中，将对文件以二进制形式操作。添加 '+' 到 mode 参数中，将允许对文件同时进行读写操作。 (1) 参数 filename：文件名称。 (2) 参数 mode：'r'（读）、'w'（写）、'a'（追加）。 (3) 参数 bufsize：如果为 0 表示不进行缓冲，如果为 1 表示进行行缓冲，如果是一个大于 1 的数表示缓冲区的大小

续表

输入输出类函数	描述
input([prompt])	获取用户输入
open(name[, mode[, buffering]])	打开文件
print	打印函数
raw_input([prompt])	设置输入，输入都是作为字符串处理

7. 自定义函数

自定义函数由用户自己创建，通常包括表示函数的关键字，用 def 表示；函数体进行一系列的逻辑计算；为函数提供数据的参数和返回值，函数执行结束后给调用者返回结果。

1) 自定义函数格式

自定义函数格式如下：

```
def 函数名(参数):
    函数体
返回值
```

示例如下：

```
def test_print():
    print('hi, python')
# 调用函数
test_print()
```

输出：hi, python

上例中函数阶段括号内没有参数，称为无参函数，在调用函数时无需传入参数。

2) 有参函数

有参函数有三种不同的参数，分别是普通参数、默认参数和动态参数。

(1) 普通参数。示例如下：

```
def test_1(varl, var2):
    print(varl+var2)
test_1(18,15)
```

以上函数中 varl、var2、18、15 是普通参数，其中 var1，var2 为形参，18，15 为实参。

(2) 默认参数。下面代码中函数 b = 0 为默认参数，如果调用函数时不传入 b 值，就使用默认值；若传入 b 值，则使用传入值。示例如下：

```
def test_2(a,b=0):
    print(a,a+b)
test_2(3)
```

输出：3 3

下面代码中函数分别传入 b = 8 和 b = 63，函数输出 a + b 分别为 14，85，示例如下：

```
test_2(6,8)
```

输出：6 14

test_2(a=22,b=63)

输出：22 85

(3) 动态参数。动态参数可以接收任意个参数，有两种动态参数，分别是 *args 和 **kwargs。使用这两种参数时 *args 必须在 **kwargs 之前；参数顺序：位置参数，*args，默认参数，**kwargs。其中 *args 接收的是按照位置传参的值，组织成一个元组；**kwargs 接收的是按照关键字传参的值，组成一个字典。动态参数使用方法有以下几种：

① *args 接受 N 个位置参数，输出结果转换成元组形式。示例如下：

```
def test_d(*args):
    print(args)
test_d(3,2,1,4,6)
```
输出：(3,2,1,4,6)
```
test_d(*[3,2,1,4,6])
```
输出：(3,2,1,4,6)

② x 接收一个值，*args 接收剩下的，并以一个元组返回。示例如下：

```
def test(x,*args):
    print(x)
    print(args)
test(1,2,6,4,7,5,8)
```
输出：1
(2, 6, 4, 7, 5, 8)

③ **kwargs 接受 N 个关键字参数，转换成字典的方式。示例如下：

```
def test(**kwargs):
    print(kwargs)
    print(kwargs['name'])
    print(kwargs['age'])
#下面两种写法执行结果一样
test(name='Bob',age=23,sex='M',birth=1999)
```
输出：{'name': 'Bob', 'age': 23, 'sex': 'M', 'birth': 1999}
Bob
23
```
test(**{'name':'Bob','age':23,'sex':'M','birth':1999})
```
输出：{'name': 'Bob', 'age': 23, 'sex': 'M', 'birth': 1999}
Bob
23

④ name 按位置接收 Amy，age 接收 14，kwargs 接收 N 个关键字，转成字典方式。示例如下：

```
def testa(name,age=18,**kwargs):
    print(name)
    print(age)
```

```
    print(kwargs)
testa('Amy',age=14,sex='F',hobby='swimming')
```
输出：Amy
14
{'sex': 'F', 'hobby': 'swimming'}

⑤ 混合使用 *args 和 **kwargs。示例如下：
```
def record(source):
    print("from %s" %   source)

def test_y(name,age=13,*args,**kwargs):
    print(name)
    print(age)
    print(args)
    print(kwargs)
    record("Hello ,testing the record now")

test_y('sara',age=100,sex='unknown',hobby='robot-making')
```
输出：
sara
100
()
{'sex': 'unknown', 'hobby': 'robot-making'}
from Hello ,testing the record now

⑥ 如果 *arg 也需要接收参数，那么前面不能有默认参数，不然调用时会报错。*kwargs 会把多传入的参数变成一个 dict 形式。示例如下：
```
def commod_record(name,history,*args,**kwargs):
    print(name)
    print(history)
    print(args)
    print(kwargs)
commod_record("Smithfield", 20, "USA","India",sex="unknown")
```
输出：
Smithfield
20
('USA', 'India')
{'sex': 'unknown'}

其中 sex="unknown" 为多传入的参数，输出结果中 sex="unknown" 变成了字典形式。

3）空函数

空函数是指一个没有包含任何执行语句的函数。它通常作为一个占位符使用，用于未来添加代码，或者在某些语境下为满足语法要求而必须声明的函数，暂时不需要实现任何功能。示例如下：

```
def func():
    Pass
```

4) return 语句

return 语句用于退出函数，选择性地向调用方返回一个表达式。如果不写 return 返回值，函数外调用将返回 None。

```
def sum(arg_a, arg_b):
    TTL = arg_a + arg_b
    print("函数内:",TTL)
    return TTL
#调用 sum 函数
TTL = sum(199, 201)
print("函数外:", TTL)
```
输出：

函数内: 400

函数外: 400

如果去除 return TTL 语句，输出如下：

函数内: 400

函数外: None

8. 匿名函数

当不需要显式定义函数时，直接传入匿名函数更方便。匿名函数用关键字 lambda 定义，只能有一个表达式，不用写 return，返回值就是该表达式的结果。定义普通方式函数、匿名函数和传入匿名函数。

(1) 普通方式函数。示例如下：

```
def com_func(var):
    return var + 1
执行函数
result = com_func(221)
print(result)
```
输出：222

(2) 匿名函数(lambda 表达式)。示例如下：

```
test_lambda = lambda com_func: com_func + 1
执行函数
result = test_lambda(221)
print(result)
```
输出：222

(3) 传入匿名函数。以 map() 函数为例，计算 $f(x) = x^2$ 时，除了定义一个 f(x) 的函数外，还可以直接传入匿名函数，示例如下：

```
tst_lam = list(map(lambda x: x * x, [4, 5, 6, 1, 3, 2, 7]))
print(tst_lam)
输出：[16, 25, 36, 1, 9, 4, 49]
```

匿名函数 lambda x: x * x 可以用普通方式函数表示如下：

```
def f(x):
    return x * x
a=[4, 5, 6, 1, 3, 2, 7]
将 a 中的元素分别带入到函数中
m=map(f,a)
print(list(m))
输出：[16, 25, 36, 1, 9, 4, 49]
```

使用 for 语句可表示如下：

```
a=[4, 5, 6, 1, 3, 2, 7]
nnpower = []
for val in a:
    nnpower.append(val*val)
print(nnpower)
```

输出结果与普通方式函数、匿名函数方法所得结果相同。

第 4 章 自然语言处理工具包

4.1 NLTK 概况及安装

1. NLTK 概况

NLTK(Natural Language Toolkit)是由宾夕法尼亚大学计算机和信息科学系的两位老师使用 Python 语言开发的一种自然语言工具包,收集了大量公开数据集、模型,提供了全面、易用的接口,涵盖了分词、词性标注(Part-Of-Speech tag,POS-tag)、命名实体识别(Named Entity Recognition,NER)、句法分析(Syntactic Parse)等各项自然语言处理(NLP)领域的功能。

NLTK 是 Python 程序处理人类语言数据的领先平台。它为 50 多个语料库和词汇资源(如 WordNet)提供了易于使用的接口,有一套用于分类、标注、提取词干、解析和推理语义的文本处理库,以及有行业优势的 NLP 库和活跃的论坛。

由于包含介绍编程基础知识以及计算语言学主题的实践指南,再加上全面的 API 文档,因此 NLTK 非常适合语言学家、工程师、学生、教育工作者、研究人员和行业用户使用。NLTK 可在 Windows、Mac OS X 和 Linux 平台上使用。最值得一提的是 NLTK 是一个免费、开源、社区驱动的项目。NLTK 被称为"使用 Python 进行计算语言学教学、工作的了不起的工具"和"神奇的自然语言处理库"。

2. 安装 NLTK

安装 NLTK 需要 Python 3.5 以上版本。对于 Windows 用户,强烈建议阅读安装指南以便安装。在 Mac、Unix 系统中安装 NLTK,可运行命令 pip Install--user-U NLTK。若需安装 Numpy(可选),可运行命令 pip Install--user-U Numpy。安装完成后,测试安装,先运行 Python,然后键入 import nltk。

1) Windows 系统 32 位中安装 NLTK

以 Python 3.8 为例,安装 NLTK 的步骤如下:

(1) 安装 Python 3.8(避免使用 64 位版本)。

(2) 安装 Numpy(可选)。

(3) 安装 NLTK。

(4) 测试安装。Windows 用户按"Win + R"键,在弹出的对话框内输入 cmd 后按"Enter"

键进入 DOS 命令行模式，输入 Python 后按"Enter"键，然后键入 import nltk。

（5）下载 NLTK 数据。安装 NLTK 包后，需下载特定功能所需的数据集/模型。如果不确定需要哪些数据集/模型，可以在命令行输入以下语句下载数据：

```
python -m nltk.downloader popular
```

或者在 Python 解释器里输入：

```
import nltk
nltk.download('popular')
```

2）在已安装 Anaconda 的 Windows 系统中安装 NLTK

在已安装 Anaconda 的 Windows 系统中安装 NLTK 的具体方法如下：

（1）若已安装 Anaconda，则 NLTK 模块已存在于终端，文件位于 Anaconda3\Lib\site-packages 里。在 DOS 命令提示符下输入 pip install nltk 后按"Enter"键，终端显示已安装 NLTK 和安装路径，界面如图 4-1 所示。

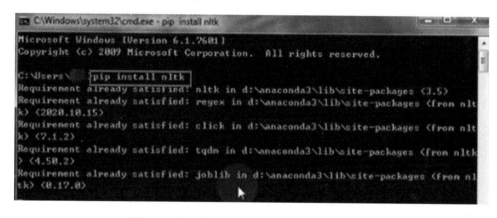

图 4-1　已安装 Anaconda 3 后安装 NLTK 的界面

（2）安装 NLTK 程序后须下载 NLTK 数据，下载步骤如下：

① 按"Win+R"键，在弹出的对话框里输入 cmd 后按"Enter"键，显示 DOS 命令提示符界面。

② 输入 Python 后按"Enter"键，界面如图 4-2 所示。

图 4-2　运行 Python 界面

③ 输入 import nltk 后按"Enter"键，再输入 nltk.download()后按"Enter"键，界面如图 4-3 所示。

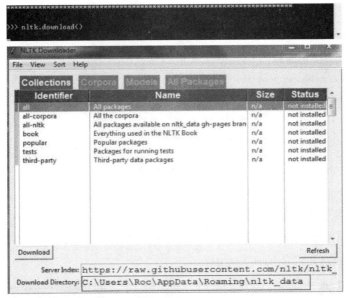

图 4-3　下载 NLTK 数据界面

也可选择修改安装路径，点击 download 下载数据，界面如图 4-4 所示。

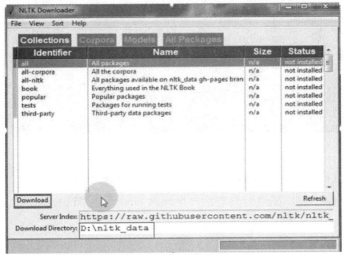

图 4-4　修改 NLTK 下载路径界面

以 Anaconda 安装路径在 D 盘根目录下为例，NLTK 的安装路径可参考图 4-5 所示界面。

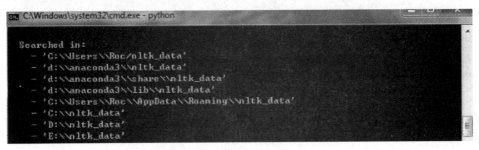

图 4-5　NLTK 安装路径界面

若下载速度较慢或报错，也可以在 C:\Windows\System32\drivers\etc 中找到 hosts 文件，在其中添加 199.232.68.133 raw.githubusercontent.com 映射后重新下载。

④ 下载完成后界面如图 4-6 所示。

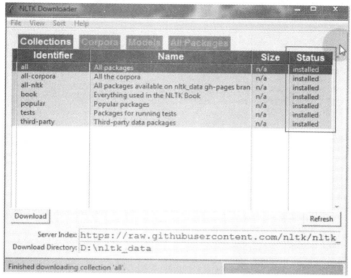

图 4-6　NLTK 数据下载完成界面

⑤ 测试 NLTK 下载是否成功。在 DOS 命令提示符下输入：

python

import nltk

from nltk.book import *

如果显示如图 4-7 所示的界面，则表明 NLTK 下载成功。

图 4-7　测试 NLTK 安装界面

如果下载 NLTK 数据速度较慢或无法下载，则可在如图 4-8 所示界面下载 NLTK 数据。

Python 在语言研究中的应用

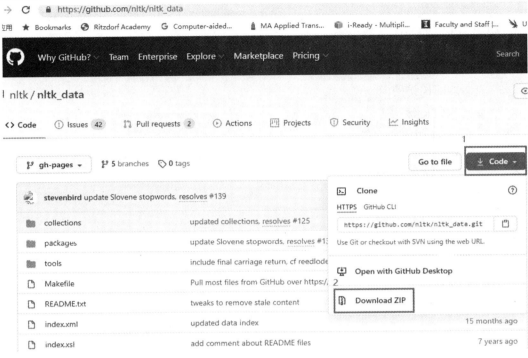

图 4-8　NLTK github 下载界面

若依旧无法下载 NLTK 数据，则可以下载网盘 NLTK 数据文件后再安装，方法如下：

a. 把网盘下载的 NLTK 数据解压，命名为 nltk_data。

b. 在 Python 提示符下输入以下命令查看安装路径：

import nltk

from nltk.book import *

c. 显示如图 4-5 所示的 NLTK 安装路径，把解压后的 nltk_data 文件夹放在其中任意安装路径下。

d. 输入 from nltk.book import*，出现如图 4-7 所示内容，说明 NLTK 数据文件下载成功。

3) 查看 NLTK 版本

使用以下命令查看 NLTK 版本：

import nltk

print(nltk.__doc__)

输出：

The Natural Language Toolkit (NLTK) is an open source Python library

for Natural Language Processing.　A free online book is available.

(If you use the library for academic research, please cite the book.)

Steven Bird, Ewan Klein, and Edward Loper (2009).

Natural Language Processing with Python.　O'Reilly Media Inc.

http://nltk.org/book

@version: 3.4.5

版本因安装时间和更新情况不同而不同。

4.2 NLTK 语料库及资源

1. 查看 NLTK 访问点及语料库

1) 查看 NLTK 访问点

查看在 NLTK 中定义了哪些访问点。代码如下：

```
import nltk
print(dir(nltk.corpus))
```

输出：

['_LazyModule__lazymodule_globals', '_LazyModule__lazymodule_import', '_LazyModule__lazymodule_init', '_LazyModule__lazymodule_loaded', '_LazyModule__lazymodule_locals', '_LazyModule__lazymodule_name', '__class__', '__delattr__', '__dict__', '__dir__', '__doc__', '__eq__', '__format__', '__ge__', '__getattr__', '__getattribute__', '__gt__', '__hash__', '__init__', '__init_subclass__', '__le__', '__lt__', '__module__', '__name__', '__ne__', '__new__', '__reduce__', '__reduce_ex__', '__repr__', '__setattr__', '__sizeof__', '__str__', '__subclasshook__', '__weakref__']

2) 查看 NLTK 语料库

查看 NLTK 中有哪些语料库。代码如下：

```
import os
import nltk
print( os.listdir( nltk.data.find("corpora") ) )
```

输出：

['abc', 'abc.zip', 'alpino', 'alpino.zip', 'biocreative_ppi', 'biocreative_ppi.zip', 'brown', 'brown.zip', 'brown_tei', 'brown_tei.zip', 'cess_cat', 'cess_cat.zip', 'cess_esp', 'cess_esp.zip', 'chat80', 'chat80.zip', 'city_database', 'city_database.zip', 'cmudict', 'cmudict.zip', 'comparative_sentences', 'comparative_sentences.zip', 'comtrans.zip', 'conll2000', 'conll2000.zip', 'conll2002', 'conll2002.zip', 'conll2007.zip', 'crubadan', 'crubadan.zip', 'dependency_treebank', 'dependency_treebank.zip', 'dolch', 'dolch.zip', 'europarl_raw', 'europarl_raw.zip', 'floresta', 'floresta.zip', 'framenet_v15', 'framenet_v15.zip', 'framenet_v17', 'framenet_v17.zip', 'gazetteers', 'gazetteers.zip', 'genesis', 'genesis.zip', 'gutenberg', 'gutenberg.zip', 'ieer', 'ieer.zip', 'inaugural', 'inaugural.zip', 'indian', 'indian.zip', 'jeita.zip', 'kimmo', 'kimmo.zip', 'knbc.zip', 'lin_thesaurus', 'lin_thesaurus.zip', 'machado.zip', 'mac_morpho', 'mac_morpho.zip', 'masc_tagged.zip', 'movie_reviews', 'movie_reviews.zip', 'mte_teip5', 'mte_teip5.zip', 'names', 'names.zip', 'nombank.1.0.zip', 'nonbreaking_prefixes', 'nonbreaking_prefixes.zip', 'nps_chat', 'nps_chat.zip', 'omw', 'omw.zip', 'opinion_lexicon', 'opinion_lexicon.zip', 'panlex_swadesh.zip', 'paradigms', 'paradigms.zip', 'pil', 'pil.zip', 'pl196x', 'pl196x.zip', 'ppattach', 'ppattach.zip', 'problem_reports', 'problem_reports.zip', 'product_reviews_1', 'product_reviews_1.zip', 'product_reviews_2', 'product_reviews_2.zip', 'propbank.zip', 'pros_cons', 'pros_cons.zip', 'ptb', 'ptb.zip', 'qc', 'qc.zip', 'reuters.zip', 'rte', 'rte.zip', 'semcor.zip', 'senseval', 'senseval.zip', 'sentence_polarity', 'sentence_polarity.zip', 'sentiwordnet', 'sentiwordnet.zip', 'shakespeare', 'shakespeare.zip', 'sinica_treebank', 'sinica_treebank.zip', 'smultron', 'smultron.zip', 'state_union', 'state_union.zip', 'stopwords', 'stopwords.zip', 'subjectivity', 'subjectivity.zip', 'swadesh', 'swadesh.zip', 'switchboard', 'switchboard.zip', 'timit', 'timit.zip', 'toolbox', 'toolbox.zip', 'treebank', 'treebank.zip', 'twitter_samples', 'twitter_samples.zip', 'udhr', 'udhr.zip', 'udhr2', 'udhr2.zip', 'unicode_samples', 'unicode_samples.zip', 'universal_treebanks_v20.zip', 'verbnet', 'verbnet.zip', 'webtext', 'webtext.zip', 'wordnet', 'wordnet.zip', 'wordnet_ic', 'wordnet_ic.zip', 'words', 'words.zip', 'ycoe', 'ycoe.zip', 'zph_corp']

3) 获取文本语料和词汇资源

(1) 获取文本语料。代码如下：

```
import nltk
from nltk.book import *
```

输出：

*** Introductory Examples for the NLTK Book ***

Loading text1, ..., text9 and sent1, ..., sent9

Type the name of the text or sentence to view it.

Type: 'texts()' or 'sents()' to list the materials.

text1: Moby Dick by Herman Melville 1851

text2: Sense and Sensibility by Jane Austen 1811

text3: The Book of Genesis

text4: Inaugural Address Corpus

text5: Chat Corpus

text6: Monty Python and the Holy Grail

text7: Wall Street Journal

text8: Personals Corpus

text9: The Man Who Was Thursday by G . K . Chesterton 1908

(2) 获取文本词汇资源。

① 查找 text1 中的 monstrous 及其上下文并只显示前 5 个结果。代码如下：

```
print(text1.concordance('monstrous,lines=5'))
```

输出：

Displaying 5 of 11 matches:

ong the former , one was of a most monstrous size . ... This came towards us ,

ON OF THE PSALMS . " Touching that monstrous bulk of the whale or ork we haver

ll over with a heathenish array of monstrous clubs and spears. Some were thick

d as you gazed , and wondered what monstrous cannibal and savage could ever hav

that has survived the flood; most monstrous and most mountainous ! That Himmal

None

② 查找 text1 中 monstrous 的同义词。代码如下：

```
print(text1.similar('monstrous'))
```

输出：

true contemptible christian abundant few part mean careful puzzled

mystifying passing curious loving wise doleful gamesome singular

delightfully perilous fearless

③ 查找 text2 中与 'monstrous'，'very' 相同的语境。代码如下：

print(text2.common_contexts(['monstrous','very']))   #相同上下文、语义场

输出：

a_pretty am_glad a_lucky is_pretty be_glad

④ 查看 text2 中 'Elinor' 'Edward' 'Marianne' 'Willoughby' 的离散分布。代码如下：

print(text2.dispersion_plot(['Elinor','Edward','Marianne','Willoughby']))

输出结果如图 4-9 所示。

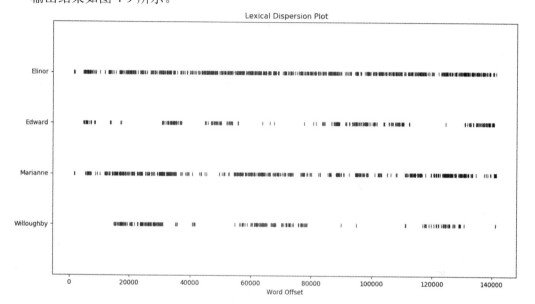

图 4-9 'Elinor' 'Edward' 'Marianne' 'Willoughby' 的离散分布图

Text2 文本转集合之后按字母顺序排序，含符号、数字和字符串：

print(sorted(set(text2)))

⑤ 查看 text2 词数。代码如下：

print(len(set(text2)))

输出：

6833

⑥ 查看 text2 中词的重复使用频率。代码如下：

print(len(text2)/len(set(text2)))

输出：

20.719449729255086

⑦ 统计 text1 中 'monstrous' 的次数。代码如下：

print(text1.count('monstrous'))

输出：

10

⑧ 统计 text1 中 'monstrous' 出现的比例。代码如下：

print(100*text1.count('monstrous')/len(text1))

输出：

0.003834076505162584

⑨ 统计 text1 词数，并生成字典。代码如下：

fdist1 =FreqDist(text1)

print(fdist1.N())

输出：

260819

⑩ 统计 text1 中出现最多的前 20 个词或字符。代码如下：

val_sort_R = sorted(fdist1.items(), key=lambda item:item[1], reverse=True)[:20]

print(val_sort_R)

输出：

[(',', 18713), ('the', 13721), ('.', 6862), ('of', 6536), ('and', 6024), ('a', 4569), ('to', 4542), (';', 4072), ('in', 3916), ('that', 2982), ('"', 2684), ('-', 2552), ('his', 2459), ('it', 2209), ('I', 2124), ('s', 1739), ('is', 1695), ('he', 1661), ('with', 1659), ('was', 1632)]

⑪ 统计 text1 词汇字典的所有键数。代码如下：

print(len(fdist1.keys()))

输出：

19317

⑫ 统计 text1 集合词汇数。代码如下：

print(len(set(text1)))

输出：

19317

⑬ 统计 text1 中出现频率最高的前 20 个词或字符并绘制频率图。代码如下：

print (fdist1.plot(20)) #频率图

输出结果如图 4-10 所示。

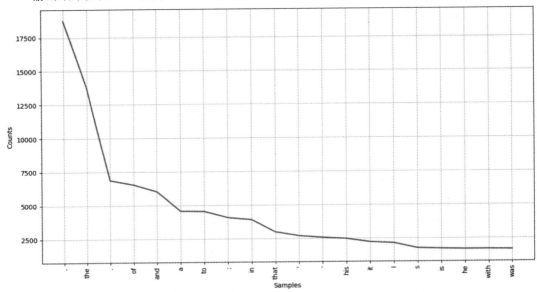

图 4-10　text1 中频率最高的前 20 个词或字符频率图

⑭ 统计 text1 中只出现一次的前五个单词，包括符号。代码如下：

print(fdist1.hapaxes()[0:5])

输出：

['Herman', 'Melville', ']', 'ETYMOLOGY', 'Late']

⑮ 统计 text1 中的 15 个字母以上的长词。代码如下：

long_words = [w for w in set(text1) if len(w)>15 and fdist1[w]>2]

print('长词数:',len(long_words),'长词:',long_words)

输出：

长词数: 3 长词: ['indiscriminately', 'apprehensiveness', 'comprehensiveness']

⑯ 对⑮中的长词排序。代码如下：

print(sorted(long_words))

输出：

['apprehensiveness', 'comprehensiveness', 'indiscriminately']

⑰ 查看 text5 中二元序列的前五个。代码如下：

from nltk.util import tokenwrap

print(tokenwrap(text5.collocation_list()[:5], separator="; "))

输出：

wanna chat; PART JOIN; MODE #14-19teens; JOIN PART; PART PART

⑱ 合并 sent1 列表元素。代码如下：

a=''.join(sent1)

print(a)

输出：

CallmeIshmael.

⑲ 切分列表元素。代码如下：

print(str(sent1).split())

输出：

["['Call',", "'me',", "'Ishmael',", "'.']"]

⑳ 查看 sent1、sent2 长度。代码如下：

print(len(sent1+sent2))

输出：

15

print(len(sent1)+len(sent2))

输出：

15

㉑ 查看 sent3 句子。代码如下：

print(sent3)

输出：

['In', 'the', 'beginning', 'God', 'created', 'the', 'heaven', 'and', 'the', 'earth', '.']

㉒ 查看 sent3 中"the"第一次出现的位置。代码如下：

print(sent3.index('the'))

输出：

1

㉓ 查看 sent3 中所有"the"的位置。代码如下：

print([i for i,v in enumerate(sent3) if v=='the'])

输出：

[1, 5, 8]

㉔ 查找 text5 中以"ru"开头的单词。代码如下：

print(sorted([w for w in text5 if w.startswith('ru')]))

输出：

['rub', 'rubs', 'rubs', 'rubs', 'ruff', 'rule', 'rule', 'rule', 'rule', 'rule', 'rule', 'rule', 'rules', 'rules', 'rules', 'rules', 'rules', 'rum', 'rum', 'rum', 'rum', 'rumour', 'rumours', 'run', 'run', 'run', 'run', 'run', 'running', 'running', 'runs', 'runs', 'runs', 'runs', 'runs', 'runs', 'runs', 'runs', 'rush']

㉕ 查看 text9 中"sunset"出现的位置。代码如下：

print(text9.index('sunset'))

输出：

629

㉖ 查看 text6 中单词长度为 12 个字母的单词。代码如下：

list4=[w for w in text6 if len(w)==12 and w.isalpha()]

freqdist4=FreqDist(list4)

sorteddict = sorted(freqdist4.items(), key=lambda item:item[1],reverse=True)

print(sorteddict)

输出：

[('BRIDGEKEEPER', 16), ('Dramatically', 2), ('bridgekeeper', 2), ('dictatorship', 1), ('perpetuating', 1), ('distributing', 1), ('elderberries', 1), ('disheartened', 1), ('individually', 1), ('accomplished', 1), ('considerable', 1), ('particularly', 1), ('intermission', 1), ('illegitimate', 1)]

㉗ 只输出单词，不输出其出现的频率，按出现频率由高到低排列。代码如下：

for item in sorteddict:

print (item[0],end="; ")

输出：

BRIDGEKEEPER; Dramatically; bridgekeeper; dictatorship; perpetuating; distributing; elderberries; disheartened; individually; accomplished; considerable; particularly; intermission; illegitimate;

2. NLTK 语料库

1) 古腾堡语料库

(1) 查看古腾堡语料库包含的语料资源。代码如下：

import nltk

print(nltk.corpus.gutenberg.fileids())

输出：

['austen-emma.txt', 'austen-persuasion.txt', 'austen-sense.txt', 'bible-kjv.txt', 'blake-poems.txt', 'bryant-stories.txt', 'burgess-busterbrown.txt', 'carroll-alice.txt', 'chesterton-ball.txt', 'chesterton-brown.txt', 'chesterton-thursday.txt', 'edgeworth-parents.txt', 'melville-moby_dick.txt', 'milton-paradise.txt', 'shakespeare-caesar.txt', 'shakespeare-hamlet.txt', 'shakespeare-macbeth.txt', 'whitman-leaves.txt']

(2) 查看古腾堡语料库语料的资源数。代码如下：

```
gfid = gutenberg.fileids()
print(len(gfid))
```

输出：

18

(3) 查看 burgess-busterbrown 文本中的前 60 个字符。代码如下：

```
import nltk
from nltk.corpus import gutenberg
raw = gutenberg.raw("burgess-busterbrown.txt")
raw60 = raw[:60]
print(raw60)
```

输出：

[The Adventures of Buster Bear by Thornton W. Burgess 1920]

(4) 查看 milton-paradise 文本中的前 5 个词汇(包括符号)。代码如下：

```
wgw = nltk.corpus.gutenberg.words('milton-paradise.txt')
gw_5 = gw[0:5]
print(gw_5)
```

输出：

['[', 'Paradise', 'Lost', 'by', 'John']

(5) 查看 milton-paradise 文本中的前 2 句(包括符号)。代码如下：

```
sents = gutenberg.sents('milton-paradise.txt')
s_2 = sents[0:2]
print(s_2)
```

输出：

[['[', 'Paradise', 'Lost', 'by', 'John', 'Milton', '1667', ']'], ['Book', 'I']]

(6) 查看 austen-sense 的文件长度。代码如下：

```
sense = nltk.corpus.gutenberg.words('austen-sense.txt')
print(len(sense))
```

输出：

141576

(7) 查看 austen-sense 文本中包含 love 的句子并只显示前 5 个结果。代码如下：

```
sense = nltk.Text(nltk.corpus.gutenberg.words('austen-sense.txt'))
sense.concordance("love")
```

输出：

Displaying 5 of 77 matches:

priety of going , and her own tender love for all her three children determine

es .'''' I believe you are right , my love ; it will be better that there shoul.

It implies everything amiable . I love him already .'''' I think you will li

sentiment of approbation inferior to love .'''' You may esteem him .'''' I have

n what it was to separate esteem and love .'' Mrs . Dashwood now took pains to

(8) 查看古腾堡文本中的原始字母(包含空格的数量、单词数和句子数)。代码如下：

```
for fileids in gutenberg.fileids():
num_chars = len(gutenberg.raw(fileids))
#文本中单词的数量
num_words = len(gutenberg.words(fileids))
#文本中句子的数量
num_sents = len(gutenberg.sents(fileids))
#将文本中的单词全部转换为小写，并去重
num_vocab = len(set([w.lower() for w in gutenberg.words(fileids)]))
print('字母数:',num_chars,'单词数:',num_words,'句子数:',num_sents,num_vocab)
```

输出：

字母数: 711215 单词数: 154883 句子数: 4250 12452

(9) 计算平均词长、平均句长、文本中每个词出现的平均次数和文本名。代码如下：

```
print('平均词长:',int(num_chars/num_words),'平均句长:',int(num_words/num_sents),'词次数:',int(num_words/num_vocab),'文本名:',fileids)
```

输出：

平均词长: 4 平均句长: 36 词次数: 12 文本名: whitman-leaves.txt

(10) 查看 austen-sense 文本中的原始字母，包含空格的数量、单词数和句子数。代码如下：

```
for fileids in 'austen-sense.txt':
    num_chars = len(gutenberg.raw('austen-sense.txt'))
    #文本中单词的数量
    num_words = len(gutenberg.words('austen-sense.txt'))
    #文本中句子的数量
    num_sents = len(gutenberg.sents('austen-sense.txt'))
    #将文本中的单词全部转换为小写，并去重
    num_vocab = len(set([w.lower() for w in gutenberg.words('austen-sense.txt')]))
print('字母数:',num_chars,'单词数:',num_words,'句子数:',num_sents,num_vocab)
```

输出：

字母数: 673022 单词数: 141576 句子数: 4999 6403

(11) 查看 austen-sense 的平均词长、平均句长、文本中每个词出现的平均次数和文本名。代码如下：

```
print('平均词长:',int(num_chars/num_words),'平均句长:',int(num_words/num_sents),'词次数:',int(num_words/num_vocab),'文本名:','austen-sense.txt')
```

输出:

平均词长: 4 平均句长: 28 词次数: 22 文本名: austen-sense.txt

2) 布朗语料库

(1) 调用布朗语料库，查看库目录。代码如下:

```
from nltk.corpus import brown
print(brown.categories())
```

输出:

['adventure', 'belles_lettres', 'editorial', 'fiction', 'government', 'hobbies', 'humor', 'learned', 'lore', 'mystery', 'news', 'religion', 'reviews', 'romance', 'science_fiction']

(2) 查看"humor"库单词。代码如下:

```
print(brown.words(categories='humor'))
```

输出:

['It', 'was', 'among', 'these', 'that', 'Hinkle', ...]

(3) 查看布朗语料库内"humor"目录下的情态动词词频。代码如下:

```
from nltk.corpus import brown
humor_text = brown.words(categories='humor')
fdist = nltk.FreqDist([w.lower() for w in humor_text])
modals = ['will','would','shall','should','can','could']
for mdls in modals:
    print(mdls + ':', fdist[mdls],end=' ')
```

输出:

will: 13 would: 56 shall: 2 should: 7 can: 17 could: 33

其中，end=' ' 表示输出时加空格在同一行显示结果，引号内可以换成其他字符或符号，比如符号"*""-"等。

(4) 统计布朗库不同文体的词频。代码如下:

```
import nltk
from nltk.corpus import brown
cfd = nltk.ConditionalFreqDist(
    (genre,word)
    for genre in brown.categories()
    for word in brown.words(categories=genre))
genres = ['editorial', 'fiction', 'government', 'humor', 'learned', 'mystery', 'news', 'religion', 'reviews', 'science_fiction']
conjunc_adv = ['therefore','then','however','otherwise']
cfd.tabulate(conditions=genres,samples=conjunc_adv)
print(cfd)
```

输出结果如图 4-11 所示。

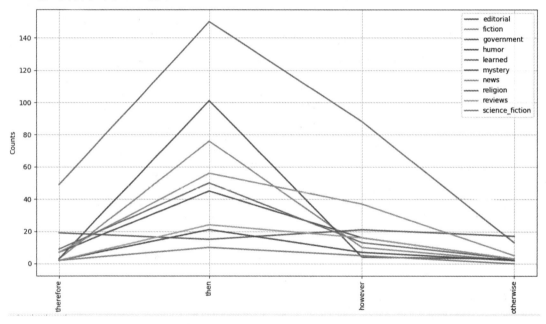

图 4-11 "therefore""then""however""otherwise"在不同文体的词频图

使用以下代码生成条件频率分布图：

```
cfd.plot(conditions=genres, samples=conjunc_adv)
```

输出结果如图 4-12 所示。

图 4-12 "therefore""then""however""otherwise"在不同文体的条件频率分布图

3) 路透社语料库

调用路透社语料库。代码如下：

```
import nltk
from nltk.corpus import reuters
```

(1) 查看路透社语料库文本编号。代码如下：

```
print(reuters.fileids())
```

输出：

['test/14826', 'test/14828', 'test/14829', 'test/14832', 'test/14833', 'test/14839', 'test/14840', 'test/14841', 'test/14842', 'test/14843', ...]

(2) 查看路透社语料库的分类。代码如下：

```
print(reuters.categories())
```

输出：

['acq', 'alum', 'barley', 'bop', 'carcass', 'castor-oil', 'cocoa', 'coconut', 'coconut-oil', 'coffee', 'copper', 'copra-cake', 'corn', 'cotton', 'cotton-oil', 'cpi', 'cpu', 'crude', 'dfl', 'dlr', 'dmk', 'earn', 'fuel', 'gas', 'gnp', 'gold', 'grain', 'groundnut', 'groundnut-oil', 'heat', 'hog', 'housing', 'income', 'instal-debt', 'interest', 'ipi', 'iron-steel', 'jet', 'jobs', 'l-cattle', 'lead', 'lei', 'lin-oil', 'livestock', 'lumber', 'meal-feed', 'money-fx', 'money-supply', 'naphtha', 'nat-gas', 'nickel', 'nkr', 'nzdlr', 'oat', 'oilseed', 'orange', 'palladium', 'palm-oil', 'palmkernel', 'pet-chem', 'platinum', 'potato', 'propane', 'rand', 'rape-oil', 'rapeseed', 'reserves', 'retail', 'rice', 'rubber', 'rye', 'ship', 'silver', 'sorghum', 'soy-meal', 'soy-oil', 'soybean', 'strategic-metal', 'sugar', 'sun-meal', 'sun-oil', 'sunseed', 'tea', 'tin', 'trade', 'veg-oil', 'wheat', 'wpi', 'yen', 'zinc']

(3) 根据文本编号查看。

① 单个文本标识作为参数。代码如下：

print(reuters.categories('test/14829'))

输出：['crude', 'nat-gas']

② 多个文本标识列表作为参数。代码如下：

print(reuters.categories(['training/9810','training/9993']))

输出：['earn']

③ 根据文本编号查看词汇。代码如下：

a = reuters.words('test/14832')

print(a)

输出：['THAI', 'TRADE', 'DEFICIT', 'WIDENS', 'IN', 'FIRST', ...]

④ 根据文本编号列表查看词汇。代码如下：

b = reuters.words(['test/14839', 'test/14840'])

print(b)

输出：['AUSTRALIAN', 'FOREIGN', 'SHIP', 'BAN', 'ENDS', 'BUT', ...]

(4) 根据分类获取词汇可分为以下两种情况。

① 根据单个分类查看词汇。代码如下：

c = reuters.words(categories='alum')

print(c)

输出：

['SHOWA', 'DENKO', 'EXPORTS', 'ALUMINIUM', 'CASTING', ...]

② 根据分类列表查看词汇。代码如下：

d = reuters.words(categories=['acq', 'barley'])

print(d)

输出：

['SUMITOMO', 'BANK', 'AIMS', 'AT', 'QUICK', 'RECOVERY', ...]

4) 美国总统就职演说语料库

调用美国总统就职演说语料库。代码如下：

import nltk

from nltk.corpus import inaugural

```
import matplotlib #调用绘图包
matplotlib.use('TkAgg')
```

(1) 查看文本编号。代码如下：

```
flids = inaugural.fileids()
print(flids)
```

输出：

['1789-Washington.txt', '1793-Washington.txt', '1797-Adams.txt', '1801-Jefferson.txt', '1805-Jefferson.txt', '1809-Madison.txt', '1813-Madison.txt', '1817-Monroe.txt', '1821-Monroe.txt', '1825-Adams.txt', '1829-Jackson.txt', '1833-Jackson.txt', '1837-VanBuren.txt', '1841-Harrison.txt', '1845-Polk.txt', '1849-Taylor.txt', '1853-Pierce.txt', '1857-Buchanan.txt', '1861-Lincoln.txt', '1865-Lincoln.txt', '1869-Grant.txt', '1873-Grant.txt', '1877-Hayes.txt', '1881-Garfield.txt', '1885-Cleveland.txt', '1889-Harrison.txt', '1893-Cleveland.txt', '1897-McKinley.txt', '1901-McKinley.txt', '1905-Roosevelt.txt', '1909-Taft.txt', '1913-Wilson.txt', '1917-Wilson.txt', '1921-Harding.txt', '1925-Coolidge.txt', '1929-Hoover.txt', '1933-Roosevelt.txt', '1937-Roosevelt.txt', '1941-Roosevelt.txt', '1945-Roosevelt.txt', '1949-Truman.txt', '1953-Eisenhower.txt', '1957-Eisenhower.txt', '1961-Kennedy.txt', '1965-Johnson.txt', '1969-Nixon.txt', '1973-Nixon.txt', '1977-Carter.txt', '1981-Reagan.txt', '1985-Reagan.txt', '1989-Bush.txt', '1993-Clinton.txt', '1997-Clinton.txt', '2001-Bush.txt', '2005-Bush.txt', '2009-Obama.txt']

(2) 按年代查看文本编号。代码如下：

```
year = [flids[:4] for flids in inaugural.fileids()]
print(year)
```

输出：

['1789', '1793', '1797', '1801', '1805', '1809', '1813', '1817', '1821', '1825', '1829', '1833', '1837', '1841', '1845', '1849', '1853', '1857', '1861', '1865', '1869', '1873', '1877', '1881', '1885', '1889', '1893', '1897', '1901', '1905', '1909', '1913', '1917', '1921', '1925', '1929', '1933', '1937', '1941', '1945', '1949', '1953', '1957', '1961', '1965', '1969', '1973', '1977', '1981', '1985', '1989', '1993', '1997', '2001', '2005', '2009']

(3) 利用条件频率显示"human""people""nation"在美国总统就职演说语料库的频率。代码如下：

```
import nltk
from nltk.corpus import inaugural
import matplotlib
matplotlib.use('TkAgg')
cfd = nltk.ConditionalFreqDist(
    (target,flids[:4])
    for flids in inaugural.fileids()
    for w in inaugural.words(flids)
    for target in ['human','people','nation']
    if w.lower().startswith(target))
e = cfd.plot()
print(e)
```

输出结果如图4-13所示。

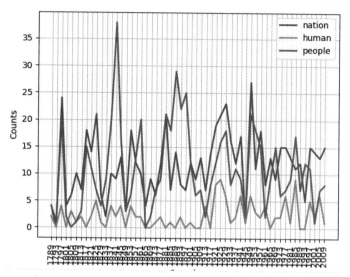

图 4-13 "human" "people" "nation" 在美国总统就职演说语料库的条件频率分布图

5) 面向语义的英语词典 WordNet

NLTK 中包括英语 WordNet 词典，是面向语义的英语词典，共有 155 287 个单词和 117 659 个同义词。WordNet 与传统辞典类似，但结构更丰富。

(1) WordNet 的层次结构。WordNet 的同义词集是抽象的概念，这些概念在层次结构中相互联系，但并不总是有对应的英语词汇。图 4-14 所示是 WordNet 概念的层次片段，每个节点对应一个同义词集，边表示上位词和下位词的关系。

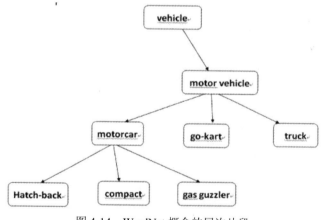

图 4-14 WordNet 概念的层次片段

(2) WordNet 的功能。调用 wordnet 包的代码如下：

```
from nltk.corpus import wordnet as wn
```

① 查询一个词所在的所有词集(synsets)。代码如下：

```
wn.synsets('car')
print(wn.synsets('car'))
```

输出：

[Synset('car.n.01'), Synset('car.n.02'), Synset('car.n.03'), Synset('car.n.04'), Synset('cable_car.n.01')]

② 查询一个同义词集的定义。代码如下：

```
defntn = wn.synset('apple.n.01').definition()
print(defntn)
```
输出：
fruit with red or yellow or green skin and sweet to tart crisp whitish flesh

③ 查询一个词语的词义。代码如下：
```
eg = wn.synset('fish.n.01').examples()
print(eg)
```
输出：
['the shark is a large fish', 'in the living room there was a tank of colorful fish']

④ 查询词语某种词性所在的同义词集合。代码如下：
```
wdsynset = wn.synsets('sheep',pos=wn.NOUN)
print(wdsynset)
```
输出：
[Synset('sheep.n.01'), Synset('sheep.n.02'), Synset('sheep.n.03')]

⑤ 查询一个同义词集中的所有词，例如，查询 car 同义词集中的所有词。代码如下：
```
allwords1 = wn.synset('car.n.01').lemma_names()
print(allwords1)
```
输出：
['car', 'auto', 'automobile', 'machine', 'motorcar']

⑥ 输出词集和词的配对——词条(lemma)。代码如下：
```
lem = wn.synset('dog.n.01').lemmas( )
print(lem)
```
输出：
[Lemma('dog.n.01.dog'), Lemma('dog.n.01.domestic_dog'), Lemma('dog.n.01.Canis_familiaris')]

⑦ 利用词条查询反义词。代码如下：
```
good = wn.synset('good.a.01')
anto = good.lemmas()[0].antonyms()
print(anto)
```
输出：
[Lemma('bad.a.01.bad')]
```
syn = wn.synsets('kind')
print(syn)
```
输出：
[Synset('kind.n.01'), Synset('kind.a.01'), Synset('kind.s.02'), Synset('kind.s.03')]
```
kind = wn.synset('kind.a.01')
anto1 = kind.lemmas()[0].antonyms()
print(anto1)
```
输出：
[Lemma('unkind.a.01.unkind')]

⑧ 查询两个词之间的语义相似度。path_similarity 函数的取值范围在 0～1 之间，值越大表示相似度越高。例如，比较羊和马的相似度。代码如下：

```
sheep = wn.synset('sheep.n.01')
horse = wn.synset('horse.n.01')
sem_sim =sheep.path_similarity(horse)
print(sem_sim)
```

输出：

0.125

比较骡和马的相似度。代码如下：

```
mule = wn.synset('mule.n.01')
horse = wn.synset('horse.n.01')
sem_sim1 = mule.path_similarity(horse)
print(sem_sim1)
```

输出：

0.3333333333333333

比较狼和狗的相似度。代码如下：

```
wolf = wn.synset('wolf.n.01')
dog = wn.synset('dog.n.01')
sem_sim2 = wolf.path_similarity(dog)
print(sem_sim2)
```

输出：

0.3333333333333333

⑨ 获取一个词的上位词。

a. 查看 motorcar 上位词的路径数。代码如下：

```
paths=motorcar.hypernym_paths()
print(len(paths))
```

输出：

2

b. 查看路径 1 的同义词集。代码如下：

```
print([synset.name() for synset in paths[0]])
```

输出：

['entity.n.01', 'physical_entity.n.01', 'object.n.01', 'whole.n.02', 'artifact.n.01', 'instrumentality.n.03', 'container.n.01', 'wheeled_vehicle.n.01', 'self-propelled_vehicle.n.01', 'motor_vehicle.n.01', 'car.n.01']

c. 查看路径 2 的同义词集。代码如下：

```
print([synset.name() for synset in paths[1]])
```

输出：

['entity.n.01', 'physical_entity.n.01', 'object.n.01', 'whole.n.02', 'artifact.n.01', 'instrumentality.n.03', 'conveyance.n.03', 'vehicle.n.01', 'wheeled_vehicle.n.01', 'self-propelled_vehicle.n.01', 'motor_vehicle.n.01', 'car.n.01']

d. 查看上位同义词集。代码如下：

print(motorcar.root_hypernyms())

输出：

[Synset('entity.n.01')]

e. 获取某个词的上位词。例如，获取"猫"的上位词。代码如下：

hyp = wn.synset("cat.n.01").hypernyms()

print(hyp)

输出：

[Synset('feline.n.01')]

获取"狗"的上位词。代码如下：

hyp1 = wn.synset("dog.n.01").hypernyms()

print(hyp1)

输出：

[Synset('canine.n.02'), Synset('domestic_animal.n.01')]

⑩ 获取一个词的下位词。

a. 查看词的下位词。例如，获取"狗"的下位词。代码如下：

hypo = wn.synset("dog.n.01").hyponyms()

print(hypo)

输出：

[Synset('basenji.n.01'), Synset('corgi.n.01'), Synset('cur.n.01'), Synset('dalmatian.n.02'), Synset('great_pyrenees.n.01'), Synset('griffon.n.02'), Synset('hunting_dog.n.01'), Synset('lapdog.n.01'), Synset('leonberg.n.01'), Synset('mexican_hairless.n.01'), Synset('newfoundland.n.01'), Synset('pooch.n.01'), Synset('poodle.n.01'), Synset('pug.n.01'), Synset('puppy.n.01'), Synset('spitz.n.01'), Synset('toy_dog.n.01'), Synset('working_dog.n.01')]

获取"车"的下位词。代码如下：

motorcar=wn.synset('car.n.01')

types_of_motorcar=motorcar.hyponyms()

print(types_of_motorcar[0])

输出：

Synset('ambulance.n.01')

b. 查看分类后"车"的同义词集。代码如下：

print(sorted(

[lemma.name()

for synset in types_of_motorcar

for lemma in synset.lemmas()]))

输出：

['Model_T', 'S.U.V.', 'SUV', 'Stanley_Steamer', 'ambulance', 'beach_waggon', 'beach_wagon', 'bus', 'cab', 'compact', 'compact_car', 'convertible', 'coupe', 'cruiser', 'electric', 'electric_automobile', 'electric_car', 'estate_car', 'gas_guzzler', 'hack', 'hardtop', 'hatchback', 'heap', 'horseless_carriage', 'hot-rod', 'hot_rod', 'jalopy', 'jeep', 'landrover', 'limo', 'limousine', 'loaner', 'minicar', 'minivan', 'pace_car', 'patrol_car', 'phaeton', 'police_car', 'police_cruiser', 'prowl_car', 'race_car', 'racer', 'racing_car', 'roadster', 'runabout', 'saloon', 'secondhand_car', 'sedan', 'sport_car', 'sport_utility', 'sport_utility_vehicle', 'sports_car', 'squad_car', 'station_waggon', 'station_wagon', 'stock_car', 'subcompact', 'subcompact_car', 'taxi', 'taxicab', 'tourer', 'touring_car', 'two-seater', 'used-car', 'waggon', 'wagon']

⑪ 获取事物可以组成的整体。代码如下：

```
part_wh = wn.synset("tree.n.01").member_holonyms()
print(part_wh)
```

输出：

[Synset('forest.n.01')]

⑫ 获取两个词在分类树中的距离。例如，获取"猫"和"狗"在分类树中的距离。代码如下：

```
cat = wn.synset("cat.n.01")
dog = wn.synset("dog.n.01")
dist = cat.shortest_path_distance(dog)
print(dist)
```

输出：

4

获取"狼"和"狗"在分类树中的距离。代码如下：

```
wolf = wn.synset('wolf.n.01')
dog = wn.synset('dog.n.01')
dist1 = wolf.shortest_path_distance(dog)
print(dist1)
```

输出：

2

⑬ 获取所有与"good"含义相似的词。代码如下：

```
good = wn.synset("good.a.1")
sim_wd = good.similar_tos()
print(sim_wd)
```

输出：

[Synset('bang-up.s.01'), Synset('good_enough.s.01'), Synset('goodish.s.01'), Synset('hot.s.15'), Synset('redeeming.s.02'), Synset('satisfactory.s.02'), Synset('solid.s.01'), Synset('superb.s.02'), Synset('well-behaved.s.01')]

⑭ 查询一个词的组成部分。代码如下：

```
pt = wn.synset('room.n.01').part_meronyms()
print(pt)
```

输出：

[Synset('ceiling.n.01'), Synset('floor.n.01'), Synset('room_light.n.01'), Synset('wall.n.01')]

⑮ 查询物质的材质。代码如下：

```
mtrl = wn.synset('wood.n.01').substance_meronyms()
print(mtrl)
```

输出：

[Synset('lignin.n.01')]

⑯ 查询词蕴含的关系。代码如下：

```
entl = wn.synset('walk.v.01').entailments()
print(entl)
```

输出：

[Synset('step.v.01')]

```
entl1 = wn.synset('eat.v.01').entailments()
print(entl1)
```

输出：

[Synset('chew.v.01'), Synset('swallow.v.01')]

3. NLTK 计算相似度

NLTK 计算相似度，输出模型的准确度、精确度、召回率和 F 值指标。代码如下：

```
from nltk.metrics import *
# 训练集和测试集
training = 'PERSON OTHER PERSON OTHER OTHER ORGANIZATION'.split()
testing = 'PERSON OTHER OTHER OTHER OTHER OTHER'.split()
# 计算并打印准确度
print("准确度：", accuracy(training, testing))
# 创建训练集和测试集的集合
trainset = set(training)
testset = set(testing)
# 计算并打印精确度、召回率和 F 值
prec = precision(trainset, testset)
rec = recall(trainset, testset)
f_val = f_measure(trainset, testset)
print("精确度、召回率和 F 值：", prec, ",", rec, ",", f_val)
```

输出：

准确度： 0.6666666666666666

精确度、召回率和 F 值： 1.0 ， 0.6666666666666666 ， 0.8

代码中准确度度量的是模型预测正确的能力，而精确度度量的是模型预测为正例的观察值中实际为正例的能力。

(1) 使用编辑距离算法应用相似性量度。代码如下：

```
print(edit_distance("relate", "relation"))
```

输出：

3

(2) 使用杰卡德系数应用相似性量度。可以定义杰卡德系数或 Tanimoto 系数为两个集合 X 和 Y 之间的相似度。代码如下：

```
X = set([10, 20, 30, 40])
Y = set([20, 30, 60])
print(jaccard_distance(X, Y))
```
输出：
0.6

(3) 使用史密斯-沃特曼算法应用相似性量度。使用 NLTK 计算玛斯距离的代码如下：

```
X = set([10, 20, 30, 40])
Y = set([30, 50, 70])
print(masi_distance(X, Y))
```
输出：
0.945

4.3 语料处理

NLTK 分词首先需要把文本分为句子，然后使用分词方法把句子切分为词汇。NLTK 分段成句使用的是 sent_tokenize 模块，调用此模块的代码如下：

```
from nltk.tokenize import sent_tokenize
```

NLTK 包括多种分词方法，以下是常用的分词包代码：

```
import nltk
from nltk.tokenize import word_tokenize
from nltk.tokenize import TreebankWordTokenizer
from nltk.tokenize import WordPunctTokenizer
from nltk.tokenize import RegexpTokenizer
from nltk.tokenize import BlanklineTokenizer
from nltk.tokenize import WhitespaceTokenizer
```

1. NLTK 分句方法

在文本分词前，通常使用 NLTK 分段成句。

(1) 使用 sent_tokenize 分句。示例代码如下：

```
import nltk
from nltk.tokenize import sent_tokenize
para = "I love Python; however, Python seems difficult for me. Therefore I strive to master it. Perhaps I can be a half programmer after learning."
print(sent_tokenize(para))
```
输出：

['I love Python; however, Python seems difficult for me.', 'Therefore I strive to master it. Perhaps I can be a half programmer after learning.']

(2) 调用 NLTK 语料库中的 webtext，用 PunktSentenceTokenizer 分句。代码如下：

```
from nltk.tokenize import PunktSentenceTokenizer
from nltk.corpus import webtext
text = webtext.raw('overheard.txt')
#使用 PunktSentenceTokenizer 进行句子分词
sent_tokenizer = PunktSentenceTokenizer()
sents1 = sent_tokenizer.tokenize(text)
```

输出全文代码如下：

```
print(str(sents1).encode('gbk','ignore').decode('gbk'))
```

输出文本中的第二句，代码如下：

```
print(sents1[1])
```

输出：

Asian girl: Yeah, being angry!

(3) 读取本地文本，用 PunktSentenceTokenizer 分句。示例代码如下：

```
from nltk.tokenize import PunktSentenceTokenizer
# 打开并读取文件
with open('filepath/filename.txt', encoding='ISO-8859-2') as f:
    text = f.read()
sent_tokenizer = PunktSentenceTokenizer(text)
sents = sent_tokenizer.tokenize(text)
# 打印句子的数量和第一个句子
print("句子的数量：", len(sents), "，第一个句子：", sents[0])
```

2. NLTK 分词方法

(1) 使用 NLTK word_tokenize 分句成词。示例代码如下：

```
from nltk.tokenize import word_tokenize
print(word_tokenize(para)) #此处使用 NLTK 分句中的变量 para
```

输出：

['I', 'love', 'Python', ';', 'however', ',', 'Python', 'seems', 'difficult', 'for', 'me', '.', 'Therefore', 'I', 'strive', 'to', 'master', 'it.Perhaps', 'I', 'can', 'be', 'a', 'half', 'programmer', 'after', 'learning', '.']

(2) N-gram 分词。二元、三元序列分词代码如下：

```
import nltk
from nltk import bigrams,trigrams
str1 = "Being thus the resultant of forces essentially revolutionary, it is not to be found in the ancient world among the material despotisms of Asia or the stationary civilisation of Egypt."
tokens = nltk.word_tokenize(str1)
# 生成并保存字符串的二元组和三元组
str_bigrams = list(nltk.bigrams(tokens))
str_trigrams = list(nltk.trigrams(tokens))
# 打印二元组
```

```
print(str_bigrams)
```
输出:

[('Being', 'thus'), ('thus', 'the'), ('the', 'resultant'), ('resultant', 'of'), ('of', 'forces'), ('forces', 'essentially'), ('essentially', 'revolutionary'), ('revolutionary', ','), (',', 'it'), ('it', 'is'), ('is', 'not'), ('not', 'to'), ('to', 'be'), ('be', 'found'), ('found', 'in'), ('in', 'the'), ('the', 'ancient'), ('ancient', 'world'), ('world', 'among'), ('among', 'the'), ('the', 'material'), ('material', 'despotisms'), ('despotisms', 'of'), ('of', 'Asia'), ('Asia', 'or'), ('or', 'the'), ('the', 'stationary'), ('stationary', 'civilisation'), ('civilisation', 'of'), ('of', 'Egypt'), ('Egypt', '.')]

```
# 打印前 5 个三元组
print("前 5 个三元组：", str_trigrams[:5])
```
输出:

前 5 个三元组: [('Being', 'thus', 'the'), ('thus', 'the', 'resultant'), ('the', 'resultant', 'of'), ('resultant', 'of', 'forces'), ('of', 'forces', 'essentially')]

n-grams 可自定义 n 值:

```
from nltk import ngrams
sentence = 'it is not to be found in the ancient world among the material despotisms of Asia or the stationary civilisation of Egypt.'
n = 6
sixgrams = ngrams(sentence.split(), n)
# 创建一个迭代器
sixgrams_iter = iter(sixgrams)
# 打印前 5 个 n-gram
for _ in range(5):
    print(next(sixgrams_iter))
```
输出:

('it', 'is', 'not', 'to', 'be', 'found')

('is', 'not', 'to', 'be', 'found', 'in')

('not', 'to', 'be', 'found', 'in', 'the')

('to', 'be', 'found', 'in', 'the', 'ancient')

('be', 'found', 'in', 'the', 'ancient', 'world')

(3) 宾州树库分词。宾州树库(Penn Treebank)是由宾夕法尼亚大学语言数据联合实验室(LDC)在 20 世纪 90 年代初开发的语料库。它包含大量标注好句法结构的英文文本，被广泛应用于句法分析、语义分析和句法语法模型训练等自然语言处理和计算语言学研究中。使用 NLTK 库进行分词，默认的分词器是基于宾州树库分词规范的 TreebankWordTokenizer。示例代码如下:

```
from nltk.tokenize import TreebankWordTokenizer
tokenizer = TreebankWordTokenizer()
print(tokenizer.tokenize(para))
```
输出:

['I', 'love', 'Python', ';', 'however', ',', 'Python', 'seems', 'difficult', 'for', 'me.', 'Therefore', 'I', 'strive', 'to', 'master', 'it.Perhaps', 'I', 'can', 'be', 'a', 'half', 'programmer', 'after', 'learning', '.']

(4) 利用 WordPunctTokenizer 分词。代码如下：

from nltk.tokenize import WordPunctTokenizer

tokenizer = WordPunctTokenizer()

print(tokenizer.tokenize(para))

输出：

['I', 'love', 'Python', ';', 'however', ',', 'Python', 'seems', 'difficult', 'for', 'me', '.', 'Therefore', 'I', 'strive', 'to', 'master', 'it', '.', 'Perhaps', 'I', 'can', 'be', 'a', 'half', 'programmer', 'after', 'learning', '.']

(5) 正则表达式分词。代码中的"\w"表示单词，示例代码如下：

from nltk.tokenize import RegexpTokenizer

tokenizer = RegexpTokenizer("[\w']+")

print(tokenizer.tokenize(para))

输出：

['I', 'love', 'Python', 'however', 'Python', 'seems', 'difficult', 'for', 'me', 'Therefore', 'I', 'strive', 'to', 'master', 'it', 'Perhaps', 'I', 'can', 'be', 'a', 'half', 'programmer', 'after', 'learning']

也可以使用以下代码代替，输出结果相同：

from nltk.tokenize import regexp_tokenize

word_tok1 = regexp_tokenize(para, "[\w']+")

print(word_tok1)

(6) BlanklineTokenizer 分词。代码如下：

from nltk.tokenize import BlanklineTokenizer

tokenizer = BlanklineTokenizer()

wd_tok_blkln = tokenizer.tokenize(para)

print(wd_tok_blkln)

输出：

['I love Python; however, Python seems difficult for me. Therefore I strive to master it. Perhaps I can be a half programmer after learning.']

该分词器根据空行来分割文本。因为在文本 para 中没有空行，所以整个文本被视为一个单独的标记，因此，对 para 分词的结果与原文相同。

(7) WhitespaceTokenizer 分词。代码如下：

from nltk.tokenize import WhitespaceTokenizer

tokenizer = WhitespaceTokenizer()

wd_tok_wts = tokenizer.tokenize(para)

print(wd_tok_wts)

输出：

['I', 'love', 'Python;', 'however,', 'Python', 'seems', 'difficult', 'for', 'me.', 'Therefore', 'I', 'strive', 'to', 'master', 'it.', 'Perhaps', 'I', 'can', 'be', 'a', 'half', 'programmer', 'after', 'learning.']

NLTK word_tokenize 或宾州树库分词器是一个强大的分词器，能处理包括标点符号在内的各种文本。它是基于宾州树库(Penn Treebank)的英语分词规则，因此对于英语文本来说效果很好。对于其他语言可能效果不佳。此外，它不能很好地处理未知词汇或者特殊词汇。

N-gram 分词可以捕捉到词语之间的上下文关系，对于某些任务来说效果很好，如语音识别、机器翻译等。N-gram 分词需要大量的文本数据来计算词语概率。由于语言的复杂性，N-gram 模型可能会遇到数据稀疏的问题。

WordPunctTokenizer 分词器可以将文本分割成单词和标点符号，对于某些需要考虑标点符号的任务来说很有用。但它只是简单将文本分割成单词和标点符号，没有考虑词语之间的上下文关系。

正则表达式分词器可以根据用户定义的规则进行分词，非常灵活。但需要用户有一定的正则表达式知识，且对复杂文本定义有效规则时可能会很困难。

BlanklineTokenizer 分词器可以将文本按空行进行分割，对于某些格式的文本(如诗歌)很有用。但它只能按空行进行分割，对于没有空行的文本效果不佳。

WhitespaceTokenizer 分词器可以将文本按空格进行分割。但它只能按空格进行分割，对于没有空格的文本或者需要考虑标点符号和上下文关系的任务，效果不理想。

3. NLTK 以外其他分词方法

1) jieba 分词

(1) jieba 分词有三种不同的分词模式：精确模式、全模式和搜索引擎模式，默认是精确模式。

① 精确模式。代码如下：

jieba.cut(sentence,cut_all=False,HMM=True)

② 全模式。代码如下：

jieba.cut(sentence,cut_all=True,HMM=True)

③ 搜索引擎模式。代码如下：

jieba.cut_for_search (sentence, HMM=True)

(2) 在对应的函数前加"l"就可以对应得到 list 结果的函数。

① 精确模式。代码如下：

jieba.lcut(sentence,lcut_all=False,HMM=True)

② 全模式。代码如下：

jieba.lcut(sentence,lcut_all=True,HMM=True)

③ 搜索引擎模式。代码如下：

jieba.lcut_for_search (sentence, HMM=True)

(3) 在精确模式，输出结果为 list 数据类型的分词。代码如下：

import jieba

sent = "I am testing the jieba cutting for its accuracy."

tst_jieba = jieba.lcut(sent,lcut_all=False,HMM=True)

在精确模式下，下述代码输出结果相同：

tst1_jieba = jieba.lcut(sent)

print(tst_jieba)

输出：

['I', ' ', 'am', ' ', 'testing', ' ', 'the', ' ', 'jieba', ' ', 'cutting', ' ', 'for', ' ', 'its', ' ', 'accuracy', '.']

在精确模式下，输出结果为 list 数据类型，对汉语分词的示例代码如下：

```
sent1 = "我今天测试了一天创建 POS 语料库,结果也没测出想要的结果,这是程序有问题,还是其他问题呢？"
sent2 = "下雨天留客天，天留人不留"
tst1_jieba = jieba.lcut(sent1,cut_all=False,HMM=True)
tst2_jieba = jieba.lcut(sent2)
print(tst1_jieba)
输出：
['我', '今天', '测试', '了', '一天', '创建', 'POS', '语料库', ',', ' ', '结果', '也', '没', '测出', '想要', '的', '结果', ',', ' ', '这是', '程序', '有', '问题', ',', ' ', '还是', '其他', '问题', '呢', '？']
print(tst2_jieba)
输出：  ['下雨天', '留客', '天', ',', ' ', '天', '留人', '不留']
```

jieba 分词是一个针对中文的分词器,能够很好地处理中文文本。它支持三种分词模式：精确模式、全模式和搜索引擎模式。此外,jieba 还支持自定义词典,可以根据需要添加新词汇。但 jieba 分词主要针对中文,对于其他语言可能效果不佳。由于 jieba 分词是基于统计的分词方法,对于一些罕见词汇或者新词,可能无法准确分词。

2) StanfordCoreNLP 分词

StanfordCoreNLP 是由斯坦福大学的 NLP 小组开源的 Java 用实现的 NLP 工具包,它对 NLP 领域的各个问题提供了解决办法。2004 年 Steve Bird 在 NLTK 中加上了对 Stanford NLP 工具包的支持,通过调用外部 jar 文件来使用 StanfordCoreNLP 工具包的功能。StanfordCoreNLP 中英文分词示例代码如下：

```
from stanfordcorenlp import StanfordCoreNLP
zh_nlp = StanfordCoreNLP(r'D:\Anaconda3\Lib\site-packages\stanford-corenlp-4.1.0', lang='zh')
sent_ZH = '西安电子科技大学位于西安，正在向世界一流奋斗。'
en_nlp = StanfordCoreNLP(r'D:\Anaconda3\Lib\site-packages\stanford-corenlp-4.1.0', lang='en')
sent_EN = "If you want to change the source code and recompile the files, see these instructions"
print ('Tokenize:', zh_nlp.word_tokenize(sent_ZH))
print ('Tokenize:', en_nlp.word_tokenize(sent_EN))
zh_nlp.close()
en_nlp.close()
输出：
Tokenize: ['西安', '电子', '科技', '大学', '位于', '西安', ',', ' ', '正', '在', '向', '世界', '一流', '奋斗', '。']
输出：
Tokenize: ['If', 'you', 'want', 'to', 'change', 'the', 'source', 'code', 'and', 'recompile', 'the', 'files', ',', 'see', 'these', 'instructions']
```

StanfordCoreNLP 是一个强大的自然语言处理工具包,它的分词器支持多种语言,包括汉语、英语、法语、德语、西班牙语等。它能够很好地处理各种复杂的文本,包括标点符号、缩写等。但 StanfordCoreNLP 需要 Java 环境运行,对于一些没有 Java 环境的用户来说会有一些不便。此外,StanfordCoreNLP 分词器的运行速度相比一些轻量级的分词器慢一些。

4. 加载停用词对文本分词

读取本地文本"overheard.txt"并使用停用词对文本分词,以文本路径 F:/PYTHONSTUDYZPH/python 2020 study/NLTK_Python 为例,代码如下:

```
with open('F:/PYTHONSTUDYZPH/python 2020 study/NLTK_Python/overheard.txt', encoding='ISO-8859-2') as f:
    text = f.read()
调用分句方法
from nltk.tokenize import sent_tokenize
加载 NLTK 语料库中的停用词表
from nltk.corpus import stopwords
english_stops = set(stopwords.words('english'))
调用分词方法
from nltk.tokenize import word_tokenize
words1 = (word_tokenize(text))
过滤停用词
words_with_stopwds1 = [word for word in words1 if word not in english_stops]
输出过滤停用词后的全部分词结果
print(words_with_stopwds1)
输出过滤停用词后的全部分词结果的前 10 个词(含字符):
print(words_with_stopwds1[:10])
['White', 'guy', ':', 'So', ',', 'plans', 'evening', '?', 'Asian', 'girl']
```

加载停用词分词时,移除这些频繁出现但对于文本主题或者含义贡献不大的词可以减少噪音,使得分词结果更加聚焦于文本的主题。

移除停用词可以减少文本的长度和词汇量,从而提高后续处理,如特征提取、模型训练等任务的效率。在某些任务如文本分类、情感分析等,移除停用词可以去除无关的信息,有助于改善模型的性能。虽然停用词通常对文本的主题贡献不大,但在某些情况下,它们也可能携带重要的信息。例如,在情感分析中,"不"这个词虽然是常见的停用词,但它对于情感的判断却非常重要。不同的任务和领域可能需要不同的停用词列表。一个好的停用词列表应该能够很好地匹配待处理文本。如果停用词列表选择不当,可能会对分词结果产生负面影响。

5. 标注

1) NLTK pos_tag 标注

NLTK pos_tag 标注示例代码如下:

```
import nltk
text = nltk.word_tokenize("And now for something completely different. This is true.")
print(nltk.pos_tag(text))
```

输出:

[('And', 'CC'), ('now', 'RB'), ('for', 'IN'), ('something', 'NN'), ('completely', 'RB'), ('different', 'JJ'), ('.', '.'), ('This', 'DT'), ('is', 'VBZ'), ('true', 'JJ'), ('.', '.')]

NLTK pos_tag 标注能够处理各种复杂的文本，包括标点符号、缩写等。但它主要针对英语，对于其他语言可能效果不佳。此外，由于它是预训练的，对于一些特定领域的文本或者新词，可能无法准确标注。

2) NLTK 默认标注

NLTK 默认标注示例如下：

```
import nltk
raw = 'I do not like green eggs and ham, I do not like them Sam I am!'
tokens = nltk.word_tokenize(raw)
default_tagger = nltk.DefaultTagger('NN')
print(default_tagger.tag(tokens))
```

输出：

[('I', 'NN'), ('do', 'NN'), ('not', 'NN'), ('like', 'NN'), ('green', 'NN'), ('eggs', 'NN'), ('and', 'NN'), ('ham', 'NN'), (',', 'NN'), ('I', 'NN'), ('do', 'NN'), ('not', 'NN'), ('like', 'NN'), ('them', 'NN'), ('Sam', 'NN'), ('I', 'NN'), ('am', 'NN'), ('!', 'NN')]

NLTK 默认标注可以为所有单词提供一个默认的词性标签，对于一些简单的任务或者作为其他标注器的后备标注器来说很有用。但由于它为所有单词提供相同的标签，因此标注结果通常不准确。

3) N-gram 标注

N-gram 标注示例如下：

```
import nltk
from nltk.corpus import treebank

test_sents = treebank.tagged_sents()[3000:]
train_sents = treebank.tagged_sents()[:3000]
from tag_util import test_para
from nltk.tag import UnigramTagger, BigramTagger, TrigramTagger
```

(1) 一元序列。示例如下：

```
unitagger = UnigramTagger(train_sents)
text1 = nltk.word_tokenize(test_para)
test1gram = unitagger.tag(text1)
print(test1gram[:10])
```

输出：

[('Soil', None), ('scientists', 'NNS'), ('are', 'VBP'), ('being', 'VBG'), ('challenged', None), ('to', 'TO'), ('provide', 'VB'), ('assessments', None), ('of', 'IN'), ('soil', None)]

```
print(unitagger.evaluate(test_sents))
```

输出：

0.8571551910209367

(2) 二元序列。示例如下：

```
bitagger = BigramTagger(train_sents)
text1 = nltk.word_tokenize(test_para)
test2gram = bitagger.tag(text1)
print(test2gram[:10])
```

输出：

[('Soil', None), ('scientists', None), ('are', None), ('being', None), ('challenged', None), ('to', None), ('provide', None), ('assessments', None), ('of', None), ('soil', None)]

```
print(bitagger.evaluate(test_sents))
```

输出：

0.11318799913662854

(3) 三元序列。示例如下：

```
tritagger = TrigramTagger(train_sents)
text1 = nltk.word_tokenize(test_para)
test3gram = tritagger.tag(text1)
print(test3gram[:10])
```

输出：

[('Soil', None), ('scientists', None), ('are', None), ('being', None), ('challenged', None), ('to', None), ('provide', None), ('assessments', None), ('of', None), ('soil', None)]

```
print(tritagger.evaluate(test_sents))
```

输出：

0.06902654867256637

(4) N 元序列组合。可以按照以下两种 N 元序列组合方式对文本进行词性标注。

① 组合 1 标注。示例如下：

```
import nltk
from tag_util import backoff_tagger
from tag_util import test_para
from nltk.corpus import treebank
from nltk.tag import UnigramTagger, BigramTagger, TrigramTagger

train_sents = treebank.tagged_sents()[:3000]
test_sents = treebank.tagged_sents()[3000:]
#创建一个默认标注器，将所有单词默认标注为名词('NN')
backoff = nltk.DefaultTagger('NN')
```

#创建后备标注器。这个标注器首先尝试使用基于三元模型的 TrigramTagger 进行标注，如果失败则使用基于二元模型的 BigramTagger，如果还失败则使用基于一元模型的 UnigramTagger，最后如果还失败则使用默认标注器。

```
ntagger = backoff_tagger(train_sents, [UnigramTagger, BigramTagger, TrigramTagger], backoff=backoff)
text1 = nltk.word_tokenize(test_para)
```

```
testngram = ntagger.tag(text1)
print(testngram[:20])
```
输出：[('Soil', 'NN'), ('scientists', 'NNS'), ('are', 'VBP'), ('being', 'VBG'), ('challenged', 'NN'), ('to', 'TO'), ('provide', 'VB'), ('assessments', 'NN'), ('of', 'IN'), ('soil', 'NN'), ('condition', 'NN'), ('from', 'IN'), ('local', 'JJ'), ('through', 'IN'), ('to', 'TO'), ('global', 'JJ'), ('scales', 'NN'), ('.', '.'), ('A', 'DT'), ('particular', 'JJ')]

```
print(ntagger.evaluate(test_sents))
```
输出：
0.8806388948845241

② 组合 2 标注。示例如下：

```
import nltk
from nltk.corpus import treebank
from tag_util import test_para
train_sents = treebank.tagged_sents()[:3000]
test_sents = treebank.tagged_sents()[3000:]
#创建默认标注器，将所有单词默认标注为名词('NN')
t0 = nltk.DefaultTagger('NN')
#创建一元标注器，并设置默认标注器为后备标注器。当一元标注器无法找到一个词的标注时，使用后默认注器。
t1 = nltk.UnigramTagger(train_sents, backoff=t0)
#创建二元标注器，并设置一元标注器为后备标注器。当二元标注器无法找到一个词的标注时，使用一元标注器。
t2 = nltk.BigramTagger(train_sents, backoff=t1)
text1 = nltk.word_tokenize(test_para)
#使用创建的二元标注器对分词后的文本进行标注
testngram = t2.tag(text1)
print(testngram[:10])
```
输出：
[('Soil', 'NN'), ('scientists', 'NNS'), ('are', 'VBP'), ('being', 'VBG'), ('challenged', 'NN'), ('to', 'TO'), ('provide', 'VB'), ('assessments', 'NN'), ('of', 'IN'), ('soil', 'NN')]

```
print(t2.evaluate(test_sents))
```
输出：
0.8810274120440319

N-gram 标注考虑词语之间的上下文关系，对于某些任务来说效果很好。但 N-gram 标注器需要大量标注过的文本数据来计算词性的概率。此外，由于语言的复杂性，N-gram 模型可能会遇到数据稀疏的问题。

4）正则表达式标注

使用自建的正则表达式对文本进行标注。示例如下：

```
import nltk
from tag_util import test_para
```

```
from nltk.tag import RegexpTagger
patterns = [ (r'^\d+$', 'CD'), (r'.*ing$', 'VBG'), (r'.*ment$', 'NN'), (r'.*ful$', 'JJ'),(r'.*ould$',
'MD'),(r'^-?[0-9]+(.[0-9]+)?$', 'CD'),(r'.*ed$', 'VBD'),(r'.*ed$', 'VBD') ]
test_sents = "I have a dream as you have dollars, commitment is very hard to make for hiring people.
requirement can't be satisfied. this could be wonderfully"
tokens = nltk.word_tokenize(test_sents)
tagger = RegexpTagger(patterns)
print(tagger.tag(tokens))
tagged_tokens = tagger.tag(tokens)
print(tagged_tokens)
```
输出：[('I', None), ('have', None), ('a', None), ('dream', None), ('as', None), ('you', None), ('have', None), ('dollars', None), (',', None), ('commitment', 'NN'), ('is', None), ('very', None), ('hard', None), ('to', None), ('make', None), ('for', None), ('hiring', 'VBG'), ('people', None), ('.', None), ('requirement', 'NN'), ('ca', None), ("n't", None), ('be', None), ('satisfied', 'VBD'), ('.', None), ('this', None), ('could', 'MD'), ('be', None), ('wonderfully', None)]

正则表达式标注根据用户定义的规则进行标注，非常灵活。但需要用户有一定的正则表达式知识，且对于复杂文本，定义有效规则可能会很困难。

5) Tnt 标注

Tnt 标注示例如下：

```
import nltk
from nltk.corpus import treebank
from tag_util import test_para
from nltk.tag import tnt

test_sents = treebank.tagged_sents()[3000:]
train_sents = treebank.tagged_sents()[:3000]
tnt_tagger = tnt.TnT(Trained=True)
tnt_tagger.train(train_sents)

text1 = nltk.word_tokenize(test_para)
tg_tokens = tnt_tagger.tag(text1)
print(tg_tokens[:20])
```
输出：
[('Soil', 'Unk'), ('scientists', 'NNS'), ('are', 'VBP'), ('being', 'VBG'), ('challenged', 'Unk'), ('to', 'TO'), ('provide', 'VB'), ('assessments', 'Unk'), ('of', 'IN'), ('soil', 'Unk'), ('condition', 'NN'), ('from', 'IN'), ('local', 'JJ'), ('through', 'IN'), ('to', 'TO'), ('global', 'JJ'), ('scales', 'Unk'), ('.', '.'), ('A', 'DT'), ('particular', 'JJ')]
```
val = tnt_tagger.evaluate(test_sents)
print(val)
```
输出：
0.875545003237643

Tnt 标注是一个基于统计的词性标注器，它使用二元和三元标注模型，能够考虑词语之间的上下文关系。然而，Tnt 标注器需要大量标注过的文本数据进行训练。此外，对于一些罕见词或者新词，可能无法准确标注。

6) 基于分类器的标注

基于分类器的标注示例代码如下：

```
import nltk
from nltk.corpus import treebank
from tag_util import test_para
from nltk.tag.sequential import ClassifierBasedPOSTagger

test_sents = treebank.tagged_sents()[3000:]
train_sents = treebank.tagged_sents()[:3000]
tagger = ClassifierBasedPOSTagger(train=train_sents)
text1 = nltk.word_tokenize(test_para)
testtnt = tagger.tag(text1)
print(testtnt[:20])
```

输出：

[('Soil', 'NNP'), ('scientists', 'NNS'), ('are', 'VBP'), ('being', 'VBG'), ('challenged', 'VBN'), ('to', 'TO'), ('provide', 'VB'), ('assessments', 'NNS'), ('of', 'IN'), ('soil', 'JJ'), ('condition', 'NN'), ('from', 'IN'), ('local', 'JJ'), ('through', 'IN'), ('to', 'TO'), ('global', 'JJ'), ('scales', 'NNS'), ('.', '.'), ('A', 'DT'), ('particular', 'JJ')]

```
print(tagger.evaluate(test_sents))
```

输出：

0.9309734513274336

NLTK 库中基于分类器的词性标注默认使用朴素贝叶斯分类器。它的优点首先表现在简单易实现，即使在某些类型数据部分缺失的情况下也能保持较好的准确性。其次，它快速且可扩展，能处理大型数据集，适用于快速学习和实时分类任务。另外，朴素贝叶斯对小型数据集表现良好，因为它不需要大量的训练数据就能产生可靠的预测。朴素贝叶斯分类器也有一些缺点。首先，它的条件独立假设可能不成立，因为在大多数情况下，特征之间显示出某种形式的依赖性。其次，在测试数据中遇到训练数据中未出现的词时，可能会出现零类概率。此外，朴素贝叶斯可能对不平衡数据的处理不佳。总的来说，朴素贝叶斯分类器是一个强大的工具，但在使用时需要注意其潜在限制。

7) MaxentClassifier 的标注

基于 MaxentClassifier 的标注示例代码如下：

```
import nltk
from nltk.corpus import treebank
from tag_util import test_para
from nltk.tag.sequential import ClassifierBasedPOSTagger
from nltk.classify import MaxentClassifier
```

```
test_sents = treebank.tagged_sents()[3000:]
train_sents = treebank.tagged_sents()[:3000]
max_tagger = ClassifierBasedPOSTagger(train=train_sents, classifier_builder=MaxentClassifier.train)
text1 = nltk.word_tokenize(test_para)
testmax = max_tagger.tag(text1)
print(testmax[:20])
```

输出：

[('Soil', 'NNP'), ('scientists', 'NNS'), ('are', 'VBP'), ('being', 'VBG'), ('challenged', 'VBN'), ('to', 'TO'), ('provide', 'VB'), ('assessments', 'NNS'), ('of', 'IN'), ('soil', 'NN'), ('condition', 'NN'), ('from', 'IN'), ('local', 'JJ'), ('through', 'IN'), ('to', 'TO'), ('global', 'JJ'), ('scales', 'NNS'), ('.', '.'), ('A', 'DT'), ('particular', 'JJ')]

```
print(max_tagger.evaluate(test_sents))
```

输出：

0.9258363911072739

MaxentClassifier 是一个最大熵分类器，它可以考虑词语的各种特征，如词形、上下文等。最大熵模型是一种灵活的模型，可以处理各种类型的数据。但训练最大熵分类器需要大量标注过的文本数据，且需要较大的计算资源。

8) Brill 标注器标注

训练并保存 Brill 标注器。示例代码如下：

```
from nltk.corpus import treebank
test_sents = treebank.tagged_sents()[3000:]

from nltk.tag import UnigramTagger
from nltk.tag import BigramTagger, TrigramTagger

from nltk.corpus import treebank
train_sents = treebank.tagged_sents()[:3000]

import numpy as np
np.set_printoptions(threshold=np.inf)

from nltk.tag import DefaultTagger
tagger = DefaultTagger('NN')

from tag_util import backoff_tagger

default_tagger = DefaultTagger('NN')
initial_tagger = backoff_tagger(train_sents, [UnigramTagger, BigramTagger, TrigramTagger], backoff=default_tagger)
initial_tagger.evaluate(test_sents)
```

```
print(initial_tagger.evaluate(test_sents))
```
输出：
0.8806388948845241

```
from tag_util import train_brill_tagger
brill_tagger = train_brill_tagger(initial_tagger, train_sents)

text1 = nltk.word_tokenize(test_para)
# print(brill_tagger.tag(text1))

brill_tagger.evaluate(test_sents)
print(brill_tagger.evaluate(test_sents))
```
输出：
0.8822361320958342

#保存 brill 标注器为 Python 支持的文件格式 Pickle，默认保存位置在当前程序文件夹。
```
import pickle
with open('brill_tagger.pickle','wb') as fw:
    pickle.dump(brill_tagger,fw)
```

#加载 brill_tagger.pickle
```
import pickle
with open('brill_tagger.pickle','rb') as fr:
    new_model = pickle.load(fr)
    from nltk.tokenize import word_tokenize
    test_sent = 'I am happy to be testing the new model again.'
    T_brill = new_model.tag(word_tokenize(test_sent))
    print(T_brill)
```
输出：
[('I', 'PRP'), ('am', 'NN'), ('happy', 'JJ'), ('to', 'TO'), ('be', 'VB'), ('testing', 'NN'), ('the', 'DT'), ('new', 'JJ'), ('model', 'NN'), ('again', 'RB'), ('.', '.')]

也可以使用以下方法调用 pickle 文件标记：
```
import nltk.data
tagger = nltk.data.load("t2.pickle")
t2_t = tagger.tag(['some', 'words', 'in', 'a', 'sentence'])
print(t2_t)
```
输出：
[('some', 'DT'), ('words', 'NNS'), ('in', 'IN'), ('a', 'DT'), ('sentence', 'NN')]

除了上述的 8 种标注方法以外，还有一些其他标注方法，如 Jieba, Textblob, spaCy

均可以用来标注，具体使用方法可以参考相关包说明使用。

Brill 标注器是基于转换规则的词性标注器，它可以自动从训练数据中学习转换规则。Brill 标注器可以很好地处理一些特定领域的文本。但 Brill 标注器需要大量标注过的文本数据进行训练。此外，由于它是基于规则的，因此对于一些不规则的词或者新词可能无法准确标注。

每种词性标注方法都有其优点和缺点，选择哪种方法需要根据具体的任务、数据和标注目的来决定。

6. 词干提取与词形还原

词干提取(Stemming)是指将词汇的附着形式去除，仅保留词干或词根的过程。词干提取结果不一定能够表达完整语义。提取过程中要去除的附着形式包括语法功能形式、时态形式、派生形式等，例如去除第三人称单数形式、去除过去时的"ed"、去除名词后缀"ment"等。NLTK 有包括阿拉伯语、葡萄牙语、德语、英语等多种语言的词干提取模块；此外，许多自然语言处理包也提供词干提取功能。

词形还原(Lemmatization)是把一个任何形式的语言词汇还原为能表达完整语义的形式，与词干提取类似，词干提取和词形还原的主流实现方法均是利用语言中存在的规则或利用词典映射提取词干或获得词的原形。

1) NLTK 词干提取

(1) nltk.stem.lancaster module 词干提取。示例代码如下：

```
from nltk.stem.lancaster import LancasterStemmer
Lanst = LancasterStemmer()
print(Lanst.stem('maximum'))
输出：'maxim'
print(Lanst.stem('presumably'))
输出：'presum'
print(Lanst.stem('multiply'))
输出： 'multiply'
print(Lanst.stem('provision'))
输出： 'provid'
print(Lanst.stem('owed'))
输出： 'ow'
print(Lanst.stem('ear'))
输出： 'ear'
print(Lanst.stem('saying'))
输出： 'say'
print(Lanst.stem('crying'))
输出： 'cry'
print(Lanst.stem('string'))
输出： 'string'
print(Lanst.stem('meant'))
```

输出：'meant'

print(Lanst.stem('cement'))

输出：'cem'

(2) nltk.stem.porter module 词干提取。示例代码如下：

from nltk.stem.porter import PorterStemmer

porter_stemmer = PorterStemmer()

print(porter_stemmer.stem('string'))

输出：string

print(porter_stemmer.stem('doing'))

输出：do

print(porter_stemmer.stem('presumably'))

输出：presum

(3) nltk.stem.regexp module 词干提取。示例代码如下：

from nltk.stem import RegexpStemmer

regst = RegexpStemmer('ing$|s$|ed$|able$|tion$|ible', min=4)

print(regst.stem('doing'))

输出：do

print(regst.stem('desks'))

输出：desk

print(regst.stem('computing'))

输出：comput

print(regst.stem('finished'))

finish

print(regst.stem('composition'))

输出：composi

print(regst.stem('possible'))

输出：poss

(4) nltk.stem.snowball module 词干提取。示例代码如下：

from nltk.stem import SnowballStemmer

snowball_stemmer = SnowballStemmer("english")

print(snowball_stemmer.stem('string'))

输出：string

print(snowball_stemmer.stem('doing'))

输出：do

print(snowball_stemmer.stem('cement'))

输出：cement

print(snowball_stemmer.stem('universities'))

输出：univers

以上四种词干提取方式总体上对规则的现在分词和过去分词处理效果较好，但在处理有不规则变化的分词还需改进。其中，第一种提取方法对这些词 "maximum" "presumably" "provision" "owed" "owed" "cement" 的提取结果不正确。第二种方法提取 "presumably" 的词干结果错误。第三种方法提取 "computing" "comosition" 的词干结果不正确。第四种提取方法提取 "universities" 的词干不正确。第一、二、四种提取方法提取器都是基于根据英语的一般规律设计启发式规则的，由于英语中有许多不规则的词，这些规则并不能覆盖所有情况，因此有时会产生错误的结果。第三种提取方法设置了去除以 ing、s、ed、able、tion 或 ible 结尾的词，并且只对长度至少为 4 的词进行词干提取，因此出现错误的词干提取结果。

2) 词形还原

(1) WordNet 词形还原。示例代码如下：

```
from nltk.stem import WordNetLemmatizer
wnl = WordNetLemmatizer()
print(wnl.lemmatize('dogs'))
输出：dog
print(wnl.lemmatize('churches'))
输出：church
print(wnl.lemmatize('aardwolves'))
输出：aardwolf
print(wnl.lemmatize('abaci'))
输出：abacus
print(wnl.lemmatize('hardrock'))
输出：hardrock
```

WordNet 词形还原是基于词典的词形还原器，它可以将词的不同形式还原为其基本形式。这种方法简单易用，对于大多数常见的词，它可以提供准确的词形还原结果。然而，由于 Wordnet Lemmatizer 默认将所有词视为名词进行处理，因此对于动词和形容词等非名词词性的词，它可能无法提供准确的词形还原结果。

(2) 标注词性 WordNet 词形还原。示例代码如下：

```
import nltk
from nltk.corpus import wordnet
from nltk.stem import WordNetLemmatizer
def get_wordnet_pos(word):
    tag = nltk.pos_tag([word])[0][1][0].upper()
    tag_dict = {"J": wordnet.ADJ,
                "N": wordnet.NOUN,
                "V": wordnet.VERB,
                "R": wordnet.ADV}
    return tag_dict.get(tag, wordnet.NOUN)

lemmatizer = WordNetLemmatizer()
```

```
#还原单词词形
word = 'dates'
print(lemmatizer.lemmatize(word, get_wordnet_pos(word)))
输出：date
#还原句子中单词词形
sentence = "there is little study on the cloud-based information services available of digital libraries"
print([lemmatizer.lemmatize(w, get_wordnet_pos(w)) for w in nltk.word_tokenize(sentence)])
输出：
['there', 'be', 'little', 'study', 'on', 'the', 'cloud-based', 'information', 'service', 'available', 'of', 'digital', 'library']
```

标注词性 WordNet 词形还原可以提供适当的词性标签。它可以根据词性将词还原为正确的基本形式。这种方法的缺点是需要先进行词性标注，这会增加处理的复杂性和时间。此外，词性标注的准确性也会影响词形还原的结果。

选择哪种词形还原方法取决于具体需求。如果需要快速简单的词形还原，可以选择 Wordnet 词形还原；如果需要更准确的词形还原，并且不介意额外的词性标注，可以选择标注词性 WordNet 词形还原。

除 NLTK 外，还有一些其他自然语言处理包可以还原词形。

3) 其他词形还原方法

(1) spaCy 词形还原。示例代码如下：

```
import spaCy
#调用 spacy 英语模型，仅保留词形还原所需的标记部分
nlp = spaCy.load('en', disable=['parser', 'ner'])
sentence = "there is little study on the cloud-based information services available of digital libraries"
#使用英语模型解析句子
doc = nlp(sentence)
#提取每个形符的词形
s_lem = "".join([token.lemma_ for token in doc])
print(s_lem)
输出：there be little study on the cloud - base information service available of digital library
```

spaCy 词形还原功能考虑词语的各种特征，如词形、上下文等。它支持汉语、英语、法语、德语、西班牙语等多种语言。它在性能上优于许多其他的自然语言处理库且还原速度较快。

(2) TextBlob 词形还原。示例代码如下：

```
from textblob import TextBlob, Word
#还原单词词形
word = 'stripes'
w = Word(word)
print(w.lemmatize())
输出：stripe
#还原句子中单词词形
```

```
sentence = "there is little study on the cloud-based information services available of digital libraries"
sent = TextBlob(sentence)
blob_lem = "". join([w.lemmatize() for w in sent.words])
print(blob_lem)
输出：there is little study on the cloud-based information service available of digital library
```

TextBlob 词形还原简单易用，它可以将词的不同形式还原为其基本形式。它的速度比 spaCy 慢，但比 NLTK 快。

(3) 标注词性的 TextBlob 词形还原。示例代码如下：

```
#定义用 POS 标记提取每个单词词形函数
from textblob import TextBlob, Word
def lemmatize_with_postag(sentence):
    sent = TextBlob(sentence)
    tag_dict = {"J": 'a',
                "N": 'n',
                "V": 'v',
                "R": 'r'}
    words_and_tags = [(w, tag_dict.get(pos[0], 'n')) for w, pos in sent.tags]
    lemmatized_list = [wd.lemmatize(tag) for wd, tag in words_and_tags]
    return "".join(lemmatized_list)
#词形还原
sentence = "There is little study on the cloud-based information services available of digital libraries"
blob_poslem = lemmatize_with_postag(sentence)
print(blob_poslem)
输出：There be little study on the cloud-based information service available of digital library
```

标注词性的 TextBlob 词形还原在进行词形还原时，提供词性标签可使词形还原结果更准确。缺点是需要先进行词性标注，这会增加处理的复杂性和时间。此外，词性标注的准确性也会影响词形还原的结果。

此外还有 Pattern Lemmatizer，Stanford CoreNLP Lemmatization，Gensim Lemmatize，TreeTagger 等，也都可以用来还原词形。总的来说，选择哪种词形还原方法取决于具体需求。如果需要快速简单的词形还原，可以选择 TextBlob 词形还原；如果需要更准确的词形还原，可以选择标注词性的 TextBlob 词形还原；如果需要一个强大的自然语言处理库，可以选择 spaCy。

4.4 文本分类

文本分类属于有监督类学习，目标是自动将文本文件分到一个或多个已定义好的类别中，文本分类主要用于分析社交媒体中的大众情感、鉴别垃圾邮件和正常邮件、自动标注客户问询和按主题分类新闻文章等。

文本分类的步骤：

(1) 准备数据集，首先加载数据并进行数据预处理，然后把数据集分为两部分：训练集和验证集。

(2) 根据特征工程训练分类器，然后建模、利用标注数据集训练机器学习模型。

特征工程包括计数向量作为特征、TF-IDF 向量作为特征(单个词语级别、多个词语级别(N-Gram)、词性级别)、词嵌入作为特征、基于文本/NLP 的特征、主题模型作为特征。分类器包括朴素贝叶斯分类器、决策树、随机梯度下降、线性分类器、随机森林算法、支持向量机(SVM)、词袋模型(Bagging Models)、助推模型(Boosting Models)、浅层神经网络和深层神经网络。深层神经网络包括卷积神经网络(CNN)、长短期记忆网络(LSTM)、门控循环单元(GRU)、双向 RNN、循环卷积神经网络(RCNN)和其他深层神经网络的变种。

1. 朴素贝叶斯分类器文本分类

(1) 利用 NLTK 提供的电影评论数据训练朴素贝叶斯分类器，提取信息最丰富的词并预测评论的情感。示例代码如下：

```python
import nltk.classify.util
from nltk.classify import NaiveBayesClassifier
from nltk.corpus import movie_reviews


def extract_features(word_list):
    return dict([(word, True) for word in word_list])
# 用 NLTK 提供的电影评论数据训练数据
if __name__ == '__main__':
    # 加载积极与消极评论
    positive_fileids = movie_reviews.fileids('pos')
    negative_fileids = movie_reviews.fileids('neg')
    features_positive = [(extract_features(movie_reviews.words(fileids=[f])), 'Positive') for f in positive_fileids]
    features_negative = [(extract_features(movie_reviews.words(fileids=[f])), 'Negative') for f in negative_fileids]
    # 数据分成训练数据集(80%)和测试数据集(20%)
    threshold_factor = 0.8
    threshold_positive = int(threshold_factor * len(features_positive))   # 800
    threshold_negative = int(threshold_factor * len(features_negative))   # 800
    # 提取特征 800 个积极文本和 800 个消极文本组成训练集，200 个积极文本和 200 个消极文本组成测试文本
    features_train = features_positive[:threshold_positive] + features_negative[:threshold_negative]
    features_test = features_positive[threshold_positive:] + features_negative[threshold_negative:]
    print("\n 训练数据点的数量:", len(features_train))
    print("测试数据点的数量:", len(features_test))
    # 训练朴素贝叶斯分类器
```

```python
        classifier = NaiveBayesClassifier.train(features_train)
        print("\n 分类器的准确性:", nltk.classify.util.accuracy(classifier, features_test))
        print("\n 十大信息最丰富的单词:")
        for item in classifier.most_informative_features()[:10]:
            print(item[0])
        # 输入简单的评论
        input_reviews = [
"The movie is heart-throbbing.",
"It is a monotonous movie. I would recommend it to no one.",
"The photography is pretty terrific in this movie.",
"The plot of the movie was terrible and the story was nothing more than my expectation."
        ]
        # 运行分类器，获得预测结果
        print("\n 预测:")
        for review in input_reviews:
            print("\n 评论:", review)
            probdist = classifier.prob_classify(extract_features(review.split()))
            pred_sentiment = probdist.max()
            # 输出
            print("预测情绪:", pred_sentiment)
            print("可能性:", round(probdist.prob(pred_sentiment), 2))
```

输出:

训练数据点的数量: 1600

测试数据点的数量: 400

分类器的准确性: 0.735

十大信息最丰富的单词:

outstanding

insulting

vulnerable

ludicrous

uninvolving

avoids

astounding

fascination

darker

symbol

预测:

评论: The movie is heart-throbbing.

预测情绪: Negative

可能性: 0.52

评论: It is a monotonous movie. I would recommend it to no one.

预测情绪: Negative

可能性: 0.7

评论: The photography is pretty terrific in this movie.

预测情绪: Positive

可能性: 0.77

评论: The plot of the movie was terrible and the story was nothing more than my expectation.

预测情绪: Negative

可能性: 0.87

上述代码使用NLTK提供的电影评论数据集进行训练和测试，无需自己收集和标注大量文本数据，利用NLTK的现成功能和数据集，实现了一个基本的文本分类流程，简单易用。只需要修改数据集和一些参数，代码就可以很容易地扩展到其他文本分类任务。

但是这种基于朴素贝叶斯的分类器在复杂的文本分类任务中，可能无法达到较高的准确率，需要更复杂的深度学习模型提升性能。代码仅使用了单词出现与否作为特征，没有进一步探索更丰富的特征，如N-Gram、词性、情感词等，这限制了模型的表达能力。代码尽管输出了最重要的10个词，但没有进一步分析这些词语在分类过程中的作用，难以解释模型的预测行为。代码实现了基本的文本分类流程，适合入门学习和快速验证。但如果要应用于更复杂的实际场景，还需要进一步优化特征工程、模型设计、超参数调优和模型解释等方面，以提高分类性能和可解释性。

(2) 利用NLTK提供的姓名作为训练集和测试集，训练朴素贝叶斯分类器，提取信息最丰富的特征，预测姓名特征及其准确性。示例代码如下：

```
import nltk
from nltk.corpus import names
import random

def gender_features(word): #特征提取器
    return {'last_letter':word[-1]} #特征集为最后一个字母 0.76

names = [(name,'male') for name in names.words('male.txt')]+[(name,'female') for name in names.words('female.txt')]
random.shuffle(names)#打乱序列

features = [(gender_features(nm),gen) for (nm,gen) in names]#返回对应的特征和标签

train,test = features[1500:],features[:300] #1500个训练集和300个测试集
classifier = nltk.NaiveBayesClassifier.train(train) #生成分类器

print('Jessy is a',classifier.classify(gender_features('Jessy')))#预测分类
```

```
print(nltk.classify.accuracy(classifier,test)) #测试准确度 该值随机变化

classifier.show_most_informative_features(10)#得到信息最丰富的10个特征
```
输出：

Jessy is a female

0.7833333333333333

Most Informative Features

```
    last_letter = 'a'           female : male  =    32.9 : 1.0
    last_letter = 'k'           male : female  =    30.2 : 1.0
    last_letter = 'f'           male : female  =    13.4 : 1.0
    last_letter = 'p'           male : female  =    10.0 : 1.0
    last_letter = 'v'           male : female  =    10.0 : 1.0
    last_letter = 'o'           male : female  =     9.4 : 1.0
    last_letter = 'm'           male : female  =     9.3 : 1.0
    last_letter = 'd'           male : female  =     8.8 : 1.0
    last_letter = 'w'           male : female  =     7.1 : 1.0
    last_letter = 'r'           male : female  =     6.2 : 1.0
```

代码使用了 NLTK 自带的人名数据集，用最后一个字母作为特征进行性别分类，这是一个简单直观的特征选择方法。由于使用了简单的特征(最后一个字母)，该分类器的预测结果具有一定的可解释性，可以方便地分析模型做出判断的依据。代码使用了 NLTK 提供的容易理解和实现的经典算法朴素贝叶斯分类器，输出了分类器在测试集上的准确率，以及信息最丰富的 10 个特征，为分类器的性能和特征重要性提供了评估。

当然，特征集仅使用最后一个字母作为特征可能无法很好地捕捉人名与性别之间的复杂关系，从而会影响分类性能。代码仅使用了 300 个测试样本，因此无法充分评估模型的泛化能力。由于随机划分训练集和测试集，导致准确率存在较大波动，这可能会影响对模型性能的判断。

(3) 利用 Sklearn 库使用贝叶斯分类电影评论。示例代码如下：

```
import pandas as pd
import numpy as np

# 加载电影评论数据集
dataset = pd.read_csv(r'filepath\imdb_review.csv')

# 将数据集划分为训练集和测试集
train_data = dataset[:35000]
test_data = dataset[35000:]

# 从训练集和测试集中提取评论文本和情感标签
```

```python
train_reviews = np.array(train_data['review'])
train_sentiments = np.array(train_data['sentiment'])
test_reviews = np.array(test_data['review'])
test_sentiments = np.array(test_data['sentiment'])

# 打印数据集大小
print(len(train_reviews),len(train_sentiments),len(test_reviews),len(test_sentiments))

# 导入文本特征提取和分类器相关的模块
from sklearn.feature_extraction.text import CountVectorizer, TfidfTransformer
from sklearn.naive_bayes import MultinomialNB
from sklearn import metrics

# 使用 CountVectorizer 将文本转换为词频向量
vect = CountVectorizer()
train = vect.fit_transform(train_reviews)

# 使用 TfidfTransformer 将词频向量转换为 TF-IDF 向量
tfidf_transformer = TfidfTransformer()
train_tfidf = tfidf_transformer.fit_transform(train)

#训练 MultinomialNB 分类器
mnb_clf = MultinomialNB().fit(train_tfidf, train_sentiments)

# 对测试集进行预处理
test = vect.transform(test_reviews)
test_tfidf = tfidf_transformer.transform(test)

# 对单个评论进行预测
review_1 = ["The movie is really impressive, but I don't like it"]
test_rev1 = vect.transform(review_1)
test_rev1_tfidf = tfidf_transformer.transform(test_rev1)
predicted = mnb_clf.predict(test_rev1_tfidf)
print(predicted)

#对整个测试集进行预测并评估模型性能
docs_test = test_tfidf
nb_predicted = mnb_clf.predict(docs_test)
accuracy = np.mean(nb_predicted == test_sentiments)
```

```
print ("The accuracy of reviews_10k is %s" %accuracy)
print(metrics.classification_report(test_sentiments, nb_predicted))
```

运行程序，结果如下：

```
35000 35000 15000 15000
['positive']
The accuracy of reviews_10k is 0.8617333333333334
```

	precision	recall	f1-score	support
negative	0.84	0.89	0.87	7490
positive	0.88	0.83	0.86	7510
accuracy			0.86	15000
macro avg	0.86	0.86	0.86	15000
weighted avg	0.86	0.86	0.86	15000

代码使用了词频和 TF-IDF 作为文本特征，并使用了 CountVectorizer 和 TfidfTransformer 对文本数据进行标准化的特征提取和转换，实现了 Scikit-learn 的基于 MultinomialNB 的文本分类任务。代码给出了分类准确率、精确率、召回率和 F1 score 等指标，有助于全面评估模型性能。这个分类模型简单且容易理解，适合初学者学习和使用。

2. 支持向量机(SVM)文本分类

利用 SVM 方法进行文本分类。示例代码如下：

```python
import pandas as pd    # 导入 pandas 库
import numpy as np     # 导入 numpy 库

# 加载电影评论数据
dataset = pd.read_csv(r'filepath\imdb_review.csv')

# 准备训练和测试数据集
train_data = dataset[:35000]    # 前 35000 条作为训练数据
test_data = dataset[35000:]     # 后面的数据作为测试数据

# 提取训练和测试数据的评论和情感标签
train_reviews = np.array(train_data['review'])
train_sentiments = np.array(train_data['sentiment'])
test_reviews = np.array(test_data['review'])
test_sentiments = np.array(test_data['sentiment'])

# 打印训练和测试数据的长度
print("评论训练集长度：", len(train_reviews),
      "情感训练集长度：", len(train_sentiments),
```

```python
    "评论测试集长度：", len(test_reviews),
    "情感测试集长度：", len(test_sentiments))

from sklearn.feature_extraction.text import CountVectorizer, TfidfTransformer   # 导入特征提取工具
from sklearn import metrics   # 导入评估工具
from sklearn.svm import SVC   # 导入支持向量机模型

# 获取词向量
vect = CountVectorizer()
train = vect.fit_transform(train_reviews)

# 获取词的 tf-idf 值
tfidf_transformer = TfidfTransformer()
train_tfidf = tfidf_transformer.fit_transform(train)
test = vect.transform(test_reviews)
test_tfidf = tfidf_transformer.transform(test)

# 使用线性核的 SVC 模型
svclf = SVC(kernel = 'linear')

# 训练模型
svclf.fit(train_tfidf,train_sentiments)

# 对测试数据进行预测
docs_test = test_tfidf
sv_predicted = svclf.predict(docs_test)

# 打印预测结果的长度
print("预测结果的长度：", len(sv_predicted))

# 将预测结果保存到 DataFrame 中
data = {"reviews":test_reviews, "sentiments":sv_predicted}
df_test = pd.DataFrame(data, columns=["reviews", "sentiments"])

# 将预测结果保存到 Excel 文件中
df_test.to_excel('svm_test_reviews_senti2.xlsx',index=False)

# 计算预测的准确率
accuracy = np.mean(sv_predicted == test_sentiments)
```

```
# 打印准确率
print ("The accuracy of reviews_classification by SVM is %s" %accuracy)

# 打印分类报告
print(metrics.classification_report(test_sentiments, sv_predicted))
```

运行程序,结果如下:

```
评论训练集长度:35000 情感训练集长度:35000 评论测试集长度 15000 情感测试集长度 15000
预测结果的长度:15000
The accuracy of reviews_classification by SVM is 0.9009333333333334
              precision    recall  f1-score   support

    negative       0.90      0.90      0.90      7490
    positive       0.90      0.90      0.90      7510

    accuracy                           0.90     15000
   macro avg       0.90      0.90      0.90     15000
weighted avg       0.90      0.90      0.90     15000
```

上述代码使用了 pandas 库来处理数据,numpy 库来进行数值计算,以及 sklearn 库进行特征提取和模型训练。为了代码的简洁性,这段代码未随机抽取数据来创建训练集和测试集,直接使用了数据集中的前 35 000 条记录作为训练数据,后面的数据作为测试数据。此外,考虑到精确度较高,没有进行代码参数调优。总的来说,这段代码提供了一个基本框架来进行文本分类任务,在实际使用时可以添加错误处理机制,改进数据划分方法,以及进行模型参数调优等。

4.5 NLTK 自定义语料库

NLTK 创建自定义语料库的步骤如下。
(1) 创建语料库数据文件夹。代码如下:

```
import os, os.path

# 定义你想要创建的目录的路径
path = os.path.join('disk_name', 'file_path', 'archive_name')
if not os.path.exists(path):
    os.mkdir(path)
print(os.path.exists(path))
```
输出:True

(2) 加载自建语料库。代码如下:

```
import nltk
```

```python
import nltk.data

#加载本地文件"NLTK_stopwords_en.txt"的内容
stopwds = nltk.data.load('NLTK_stopwords_en.txt')

# 加载本地文件"NLTK_stopwords_en.txt",设置format参数为'raw'获取文件原始的二进制内容
stopwds_t = nltk.data.load('NLTK_stopwords_en.txt',format='raw')

#使用split函数将字符串分割成单词列表,打印出前10个单词
print(stopwds.split()[0:10],'\n',stopwds_t.split()[0:10])
```
输出:
["['arabic',", "'azerbaijani',", "'danish',", "'dutch',", "'english',", "'finnish',", "'french',", "'german',", "'greek',", "'hungarian',"]
[b"['arabic',", b"'azerbaijani',", b"'danish',", b"'dutch',", b"'english',", b"'finnish',", b"'french',", b"'german',", b"'greek',", b"'hungarian',"]

(3) WordListCorpusReader 读取 CSV 或单个词语料库。代码如下:

```python
from nltk.corpus.reader import WordListCorpusReader

#读取本地文件"ph_wordlist.txt"
reader = WordListCorpusReader('.', ['zph_wordlist.txt'])
print(reader.words())
```
输出:
['oh', 'my', 'gosh', 'you', 'are', 'the', 'greatest', 'creature', 'i', 'have ', 'ever', 'seen']
```python
print(reader.fileids())
```
输出:['ph_wordlist.txt']

(4) WordListCorpusReader 读取文本内容。代码如下:

```python
import nltk.data
from nltk.corpus.reader import WordListCorpusReader

#读取本地文件"HuckleberryFinn.txt"
huck = WordListCorpusReader('.',['file_path/HuckleberryFinn.txt'])
print(huck)

# 打印文件中前3句
print(huck.words()[:3])
```
输出:
['The Project Gutenberg EBook of Adventures of Huckleberry Finn, Complete', 'by Mark Twain (Samuel Clemens)', 'This eBook is for the use of anyone anywhere at no cost and with almost']

(5) TaggedCorpusReader 读取自定义 POS 文本。代码如下:

```python
from nltk.corpus.reader import TaggedCorpusReader
```

```
# reader = TaggedCorpusReader('.', r'.pos_eg.pos')

#读取本地文件"pos_eg.pos"
reader = TaggedCorpusReader('.', r'file_path/pos_eg.pos')
tok = (reader.words('pos_eg.pos'))
print(tok)
```
输出：['based_VVN', 'on_IN', 'rna_NN', 'sequence_NN', ...]

(6) PlaintextCorpusReader 创建自定义语料库并读取。代码如下：
```
from nltk.corpus import PlaintextCorpusReader

# 定义想要读取文件的目录
root = r"archive_path\archive_name"
# 创建 PlaintextCorpusReader 对象，读取目录下的所有 txt 文件
my_corpus = PlaintextCorpusReader(root, '.*\.txt')
```
① 读取原始文本。代码如下：
```
str = my_corpus.raw('Great_Expectations.txt')
print(str[:69])
```
输出：The Project Gutenberg EBook of Great Expectations, by Charles Dickens

② 读取文件标识。代码如下：
```
print(my_corpus.fileids())
```
输出：['Great_Expectations.txt', 'HuckleberryFinn.txt', 'Syndrome by Thomas Hoover.txt']
```
print(my_corpus.words('Great_Expectations.txt'))
```
输出：['The', 'Project', 'Gutenberg', 'EBook', 'of', 'Great', ...]

③ 读取句子。代码如下：
```
sents = my_corpus.sents('Great_Expectations.txt')
print(sents)
```
输出：[['The', 'Project', 'Gutenberg', 'EBook', 'of', 'Great', 'Expectations', ',', 'by', 'Charles', 'Dickens'], ['This', 'eBook', 'is', 'for', 'the', 'use', 'of', 'anyone', 'anywhere', 'at', 'no', 'cost', 'and', 'with', 'almost', 'no', 'restrictions', 'whatsoever', '.'], ...]

④ 读取段落，以读取"Great_Expectations.txt"文本中第一段第一句为例。代码如下：
```
paras = my_corpus.paras('Great_Expectations.txt')
#打印第一段第一句
print(' '.join(paras[0][0]))
```
输出：The Project Gutenberg EBook of Great Expectations , by Charles Dickens

4.6 文本特征统计

文本特征统计步骤如下。
(1) 构造对象。代码如下：
```
import nltk
```

```
from nltk.corpus import PlaintextCorpusReader
# 指定包含文本文件的目录
root = r"archive_path\archive_name"
my_corpus = PlaintextCorpusReader(root, '.*\.txt')
text_great = nltk.text.Text(my_corpus.words('Great_Expectations.txt'))
print(text_great)
```
输出：<Text: The Project Gutenberg EBook of Great Expectations ,...>

(2) 查找单词，显示"great"出现的上下文。代码如下：

```
print(text_great.concordance('great', 50, 10))
```
输出：Displaying 10 of 141 matches:
ct Gutenberg EBook of Great Expectations , by Cha
tenberg . org Title : Great Expectations Author :
OJECT GUTENBERG EBOOK GREAT EXPECTATIONS *** Prod
n Anonymous Volunteer GREAT EXPECTATIONS [1867 E
coarse gray , with a great iron on his leg . A m
present moment , with great difficulty . I find i
s sore feet among the great stones dropped into t
and had established a great reputation with herse
round , and looked in great depression at the fir
 me up there with his great leg . " Where have yo
None

(3) 查找搭配，显示5个。代码如下：

```
from nltk.util import tokenwrap
print(tokenwrap(text_great.collocation_list()[:5], separator="; "))
```
输出：Miss Havisham; said Joe; dear boy; Jolly Bargemen; Miss Skiffins

(4) 查找相似词。代码如下：

```
print(text_great.similar('great', 5))
```
输出：good own strange this with
```
print(text_great.similar('great'))
```
输出：good own strange this with that little in your strong if hard much same bad right left large common the

(5) 查找'pretty' 'very'相同上下文。代码如下：

```
print(text_great.common_contexts(['pretty', 'very']))
```
输出：a_large

(6) 绘制'sister' 'mother' 'man' 'sir' 'I'分布图。代码如下：

```
import matplotlib
matplotlib.use('TkAgg')
print(text_great.dispersion_plot(['sister', 'mother', 'man', 'sir', 'I']))
```

输出结果如图 4-15 所示。

第 4 章 自然语言处理工具包 141

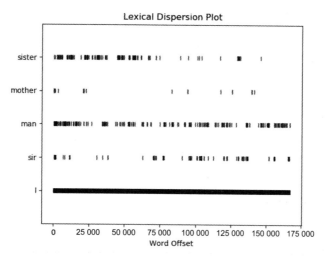

图 4-15 'sister' 'mother' 'man' 'sir' 'I' 分布图结果

(7) 用正则表达式匹配单词 "man"。代码如下：

```
from io import StringIO
import sys

# 指定搜索模式
pattern = "<.*><.*><man>"

# 获取 findall 方法的输出
old_stdout = sys.stdout    # 保存当前的 stdout 流
sys.stdout = StringIO()    # 重定向 stdout 到 StringIO 对象以获取输出
text_great.findall(pattern)    # 执行查找
output = sys.stdout.getvalue()    # 获取的输出
sys.stdout = old_stdout    # 恢复原始 stdout 流

# 处理输出，移除多余空白并分割成列表
results = [result.strip() for result in output.split(';') if result.strip()]

# 如果搜索结果超过 10 个，仅保留前 10 个
if len(results) > 10:
    results = results[:10]

# 使用逗号连接结果，并打印
print(', '.join(results))
```

输出：
, dark man, as a man, A fearful man, . A man, . A man, said the man, said the man, said the man, . The man, said the man

(8) 统计词出现的次数，如统计 "man" 的次数。代码如下：

print(text_great.count('man'))

输出：243

(9) 查看词索引。代码如下：

print(text_great.index('man'))

输出：301

(10) 查看文本总长度。代码如下：

print(len(text_great))

输出：166898

生成 FreqDist 类

dist = text_great.vocab()

print(dist)

输出：<FreqDist with 10286 samples and 166898 outcomes>

查看文中多个代词'I','you','he','we','they','we'频率

fdist = nltk.FreqDist([w.lower() for w in text_great])

prons = ['I','you','he','we','they','we']

for pron in prons:

    print(pron + ':', fdist[pron],end=' ')

输出：I: 0 you: 1673 he: 1662 we: 526 they: 283 we: 526

(11) 绘制高频词汇。代码如下：

fdist1 = nltk.FreqDist([w.lower() for w in text_great])

\# print(type(fdist1)) #<class 'nltk.probability.FreqDist'>

for key in fdist1:

    # print(key, fdist1[key])

    import matplotlib,numpy

    matplotlib.use('TkAgg')

pic1 = fdist1.plot(10)   ### 绘制高频词汇

pic1.show()

输出结果如图 4-16 所示。

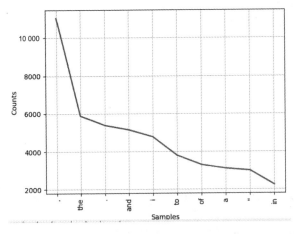

图 4-16　高频词汇图

(12) 绘制累积频数图。代码如下：

pic2 = fdist1.plot(10,cumulative=True)
print(pic2)

输出结果如图 4-17 所示。

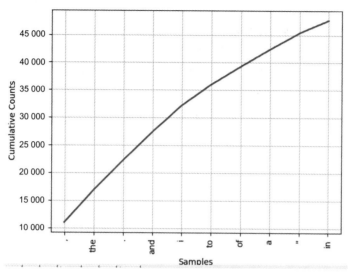

图 4-17 累积频数图

(13) 绘制出现次数最多的前 n 项列表。代码如下：

mcwds = fdist1.most_common(15)
print(mcwds)

输出：[(',', 11050), ('the', 5880), ('.', 5382), ('and', 5147), ('i', 4773), ('to', 3794), ('of', 3294), ('a', 3092), ('"', 2984), ('in', 2232), ('that', 2194), ('it', 2054), ('was', 2034), ("'", 2027), ('you', 1673)]

(14) 以表格的方式打印出现次数最多的前 n 项。代码如下：

tabwds = fdist1.tabulate(15)
print(tabwds)

输出如图 4-18 所示。

,	the	.	and	i	to	of	a	"	in	that	it	was	'	you
11050	5880	5382	5147	4773	3794	3294	3092	2984	2232	2194	2054	2034	2027	1673

图 4-18 出现次数最多的前 n 项

(15) 查找出现次数最多的项。代码如下：

maxItms = fdist1.max()
print(maxItms)

输出：,

(16) 其他常用方法及作用的代码如下：

Text(words)：构造对象。
concordance(word, width=79, lines=25)：显示 word 出现的上下文。
common_contexts(words)：显示 words 出现的相同模式。

similar(word)：显示 word 的相似词。

collocations(num=20, window_size=2)：显示最常见的二词搭配。

count(word)：word 出现的词数。

dispersion_plot(words)：绘制 words 中文档中出现的位置图。

vocab()：返回文章去重的词典。

nltk.text.TextCollection 类是 Text 的集合，提供下列方法：

nltk.text.TextCollection([text1,text2,])：构造对象。

idf(term)： 计算词 term 在语料库中的逆文档频率，即 tf(term,text) #统计 term 在 text 中的词频。

Tf-idf(term,text)：计算 term 在句子中的 tf-idf,即 tf*idf。

第 5 章 句 法 分 析

分析句法结构对于消除歧义，正确深入理解自然语言至关重要。计算机句法分析是通过文法分析和算法实现的，文法是描述语言结构的形式规则，算法是用数学表达式解析语言的方法。文法主要有两种，分别是短语结构文法和依存文法。短语结构文法是通过固定数量的规则把句子分解为短语和单词，再把短语分解为更短的短语、单词等更小的单元。依存文法是假设文法结构本质上包含词和词之间的依存(修饰)关系，一个依存关系连接两个词，分别是核心词(Head)和依存词(Dependent)，依存关系就是核心词与它的依存词之间的二元对称关系。依存关系可以细分为不同的类型，以表示两个词之间的具体句法关系。句法分析常用的自然语言处理工具包有 NLTK、spaCy、StandfordCoreNLP 和 HanLP。本章仅讲述前三个工具包的使用。

5.1 NLTK 句法分析

歧义是指语言中存在有多种可能解释的表达方式，有普遍存在的歧义和特定语境的歧义。普遍存在的歧义(Universal Ambiguity)指的是存在于语言本身的一种特性，而非特定语境或语境中的歧义。例如：She saw the man with the telescope. 此句可理解为"她用望远镜看到了那个男人"或"她看到了一个拿望远镜的男人"。这种句法歧义主要是因为语法结构的灵活性而导致的。其他普遍存在的歧义还包括由语言符号的多重含义而产生的词义歧义、由语义结构关系的复杂性而产生的语义结构歧义。特定语境的歧义由特定语境、文化背景或个人经验等而产生语用歧义。在具体的语境中，通常可以通过上下文、语用等因素来消除歧义，使得交际顺利进行。本节不研究词汇、语义结构和语用歧义，仅研究句法类歧义。普遍存在的歧义是语言学研究的重点之一，使用计算机分析句法有助于我们更好地认识和理解语言的特性。

使用 NLTK 分析句子 He shot an elephant in his pajamas 中短语 in his pajamas 的歧义，示例代码如下：

```
import nltk
CF_grammar = nltk.CFG.fromstring("""
    S -> NP VP
    PP -> P NP
```

```
    NP -> Det N | Det N PP | 'He'
    VP -> V NP | VP PP
    Det -> 'an' | 'his'
    N -> 'elephant' | 'pajamas'
    V -> 'shot'
    P -> 'in'
    """)
sent = ['He', 'shot', 'an', 'elephant', 'in', 'his', 'pajamas']
parser = nltk.ChartParser(CF_grammar)
for tree in parser.parse(sent):
    print(tree)
    tree.draw()
```

输出如下：

```
(S
  (NP He)
  (VP
    (VP (V shot) (NP (Det an) (N elephant)))
    (PP (P in) (NP (Det his) (N pajamas)))))
```

输出结果如图 5-1 所示。

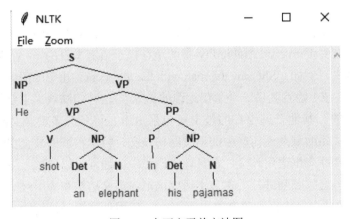

图 5-1　上下文无关文法图

正确理解上例中的句子取决于介词短语 in his pajamas 是修饰大象还是修饰句子主语 He，利用上下文无关文法图可以直观地了解句子中各组成部分的修饰关系。

1. 上下文无关文法

上下文无关文法(Context Free Grammar，CFG)是基于规则的短语结构文法，其中常用的递归下降分析法是一种最简单的分析方法。递归下降分析法将一个文法从一个高层次的目标分解成几个低层次的子目标。此外，上下文无关文法还包括移进-归约分析器和概率上下文无关文法。

(1) 递归下降分析器。递归下降分析法在解析过程中将建立分析树，首先找到一个主语，创建主语根节点。随着解析过程使用文法进行产生式递归扩展，分析树不断向下延伸，因此该算法也被命名为递归下降分析器。示例代码如下：

```
import nltk
from nltk import CFG

grammar1 = CFG.fromstring("""
s -> NP VP
VP -> V NP | V NP PP
PP -> P NP
V ->"saw" | "ate" | "walked"
NP ->"John" | "Mary" | "Bob" | Det N | Det N PP
Det ->"a" | "an" | "the" | "my"
N ->"man" | "dog" | "cat" | "telescope" | "park"
P ->"in" | "on" | "by" | "with"
""")
sent = "Mary saw Bob with a dog".split()
rd_parser = nltk.RecursiveDescentParser(grammar1)
for tree in rd_parser.parse(sent):
    print(tree)
```

输出：

(s
　　(NP Mary)
　　(VP (V saw) (NP Bob) (PP (P with) (NP (Det a) (N dog)))))

输出结果如图 5-2 所示。

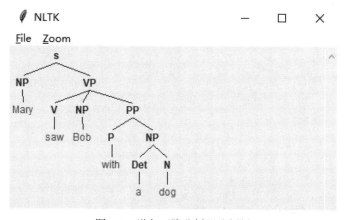

图 5-2　递归下降分析器分析树

(2) 移进-归约分析器。移进-归约分析器是自下而上的分析器，与所有自下而上的分析器一样，移进-归约分析器尝试找到对应文法产生式右侧的词和短语序列，并用左侧的替换它们，直到整个句子归约为一个 S。示例代码如下：

```
import nltk
from nltk import CFG


grammar1 = CFG.fromstring("""
s -> NP VP
VP -> V NP | V NP PP
PP -> P NP
V ->"saw" | "ate" | "walked"
NP ->"John" | "Mary" | "Bob" | Det N | Det N PP
Det ->"a" | "an" | "the" | "my"
N ->"man" | "dog" | "cat" | "telescope" | "park"
P ->"in" | "on" | "by" | "with"
""")

sr_parse = nltk.ShiftReduceParser(grammar1,trace = 2)
sent = 'Mary saw a cat'.split()
for tree in sr_parse.parse(sent):
    print(tree)
```

输出：
```
Parsing 'Mary saw a cat'
    [ * Mary saw a cat]
  S [ 'Mary' * saw a cat]
  R [ NP * saw a cat]
  S [ NP 'saw' * a cat]
  R [ NP V * a cat]
  S [ NP V 'a' * cat]
  R [ NP V Det * cat]
  S [ NP V Det 'cat' * ]
  R [ NP V Det N * ]
  R [ NP V NP * ]
  R [ NP VP * ]
  R [ s * ]
(s (NP Mary) (VP (V saw) (NP (Det a) (N cat))))
    tree.draw()
```

输出树图结果如图 5-3 所示。

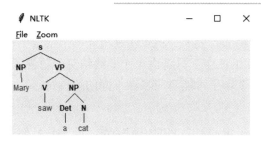

图 5-3　移位-规约分析器树图

移位-规约分析器把输入移到一个堆栈中，并尝试匹配堆栈顶部的项目和文法产生式右边的部分。这个分析器不一定保证能为输入找到一个有效的解析，即使它确实存在，也只是建立子结构而不检查它是否与全部文法一致。

(3) 概率上下文无关文法。概率上下文无关文法(Probabilistic Context-Free Grammar, PCFG)是一种上下文无关文法，它的每一个产生式关联一个概率。它会产生与相应上下文无关文法相同的文本解析，并给每个解析分配一个概率。示例代码如下：

```
PCF_grammar = nltk.PCFG.fromstring("""
    S    -> NP VP           [1.0]
    VP   -> TV NP           [0.4]
    VP   -> IV              [0.3]
    VP   -> DatV NP NP      [0.3]
    TV   -> 'saw'           [1.0]
    IV   -> 'ate'           [1.0]
    DatV -> 'gave'          [1.0]
    NP   -> 'telescopes'    [0.8]
    NP   -> 'Jack'          [0.2]
""")
# print(PCF_grammar)

viterbi_parser = nltk.ViterbiParser(PCF_grammar)
for tree in viterbi_parser.parse(['Jack', 'saw', 'telescopes']):
    print(tree)
输出：(S (NP Jack) (VP (TV saw) (NP telescopes))) (p=0.064)
    tree.draw()
```

输出树图结果如图 5-4 所示。

图 5-4　概率上下文无关文法树图

CFG 的规则清晰明了，易于理解和实现。它有足够强的表达力来表示大多数程序设计语言的语法。使用者可以构造有效的分析算法来检验一个给定字符串是否是由某个上下文无关文法产生的。但 CFG 的产生式只考虑单个非终结符，忽视了非终结符的上下文信息。它可能会产生歧义，即一个句子可能对应多棵不同的解析树。

2. 依存文法

依存文法主要关注句子中词汇之间的依赖关系，而不是具体的短语结构。它能够更好地处理复杂的句子结构和语义关系，尤其是那些成分句法分析难以处理的复杂情况。总体上，依存文法可以很好地解决歧义问题，更好地反映自然语言的表达方式。然而，依存文法需要更复杂的算法来处理各种依赖关系。某些特殊情况下它的处理结果可能不够准确。

使用依存文法分析句子，示例代码如下：

```
import nltk
gro_dep_grammar = nltk.DependencyGrammar.fromstring("""
    'shot' -> 'I' | 'elephant' | 'in'
    'elephant' -> 'an' | 'in'
    'in' -> 'pajamas'
    'pajamas' -> 'my'
    """)
pdp = nltk.ProjectiveDependencyParser(gro_dep_grammar)
sent = 'I shot an elephant in my pajamas'.split()
trees = pdp.parse(sent)
for tree in trees:
    print(tree)
输出：(shot I (elephant an (in (pajamas my))))
    tree.draw()
```

输出依存关系树图结果如图 5-5 所示。

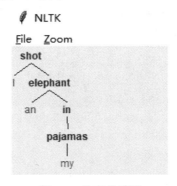

图 5-5　依存关系图

3. 利用 NLTK 树库开发文法

Nltk.corpus 模块定义了树库语料的阅读器，其中包含了宾州树库语料 10%的样本。可以查看树库文件标识，根据标识输出文件中的句子。按照句子标记规则来帮助开发文法，示例代码如下：

```
import nltk
from nltk.corpus import treebank
print(treebank.fileids())
输出：['wsj_0001.mrg', 'wsj_0002.mrg', 'wsj_0003.mrg', 'wsj_0004.mrg', 'wsj_0005.mrg', 'wsj_0006.mrg', 'wsj_0007.mrg', 'wsj_0008.mrg', 'wsj_0009.mrg', …'wsj_0199.mrg']
t = treebank.parsed_sents('wsj_0002.mrg')[0]
print(t)
输出：(S
  (NP-SBJ-1
    (NP (NNP Rudolph) (NNP Agnew))
    (, ,)
    (UCP
      (ADJP (NP (CD 55) (NNS years)) (JJ old))
      (CC and)
      (NP
        (NP (JJ former) (NN chairman))
        (PP
          (IN of)
          (NP (NNP Consolidated) (NNP Gold) (NNP Fields) (NNP PLC)))))
    (, ,))
  (VP
    (VBD was)
    (VP
      (VBN named)
      (S
        (NP-SBJ (-NONE- *-1))
        (NP-PRD
          (NP (DT a) (JJ nonexecutive) (NN director))
          (PP
            (IN of)
            (NP
              (DT this)
              (JJ British)
              (JJ industrial)
              (NN conglomerate)))))))
  (. .))
```

搜索树库找出句子的补语。代码如下：

```
def filter(tree):
    child_nodes = [child.label() for child in tree if isinstance(child, nltk.Tree)]
    return (tree.label() == 'VP') and ('S' in child_nodes)
from nltk.corpus import treebank
print([subtree for tree in treebank.parsed_sents() for subtree in tree.subtrees(filter)][5])
```

输出：(VP
  (VBZ appears)
  (S
    (NP-SBJ (-NONE- *-1))
    (VP
      (TO to)
      (VP
        (VB be)
        (NP-PRD
          (NP (DT the) (JJS highest))
          (PP
            (IN for)
            (NP
              (NP (DT any) (NN asbestos) (NNS workers))
              (RRC
                (VP
                  (VBN studied)
                  (NP (-NONE- *))
                  (PP-LOC
                    (IN in)
                    (NP
                      (JJ Western)
                      (VBN industrialized)
                      (NNS countries)))))))))))))

NLTK 语料库也收集了中国台湾的"中央研究院"现代汉语树库语料，包括 10 000 句已分析的句子。"中央研究院"现代汉语树库语料(Sinica Corpus)包含一千多万词的带标记平衡语料。语料主题包括 1981—2007 年之间的文学、生活、社会、科学、哲学和艺术文章。以下代码加载并显示这个语料库中的第 3501 个句子的句法结构树图：

```
print(nltk.corpus.sinica_treebank.parsed_sents()[3500].draw())
```

输出树图结果如图 5-6 所示。

利用 NLTK 树库开发文法句法分析具有开源免费、丰富语料库和工具、可扩展性强、易于集成、活跃社区支持等优点，但也存在性能可能不如专业工具、需要编程能力、数据预处理繁琐、语言支持有限、更新滞后等缺点。总体而言，NLTK 适合用于研究、教学和小型项目，对于大规模生产环境，需使用更专业、更高性能的商业工具。

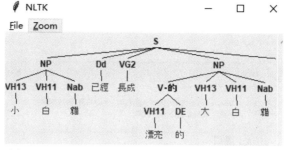

图 5-6　现代汉语树库语料库第 3501 个句法结构树图

5.2　Stanford CoreNLP 句法分析

Stanford CoreNLP 是开源的 Java 文本分析工具，包括分词器(Word Segmenter)、词性标注工具(Part-Of-Speech Tagger)、命名实体识别工具(Named Entity Recognizer)、句法分析器(Parser)等。NLTK 已经提供了相应的文本处理工具接口，包括词性标注，命名实体识别和句法分析器的接口，但 Standford CoreNLP 需安装后才可以使用。

1. 安装 Standford CoreNLP

安装 Standford CoreNLP 的步骤如下：

(1) 在官网下载安装程序，如果系统没有 Java，则先安装 Java JDK，安装后需配置 JAVA-HOME 和 path，例如将 Java 安装在 D 盘，新建变量界面如图 5-7 所示，Java 程序路径如图 5-8 所示。

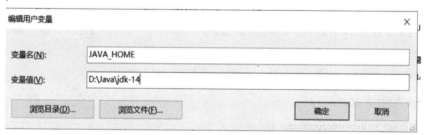

图 5-7　JAVA-HOME 新建变量界面

图 5-8　Java 程序路径

检查 Java 程序安装是否成功，使用 WIN + R 调出"运行"，输入"cmd"后按"Enter"键进入 DOS 状态，界面如图 5-9 所示。

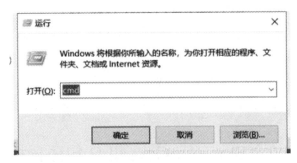

图 5-9　进入 DOS 界面

在提示符下输入"java –version"后按"Enter"键，显示已安装 Java 程序版本等信息，界面如图 5-10 所示。

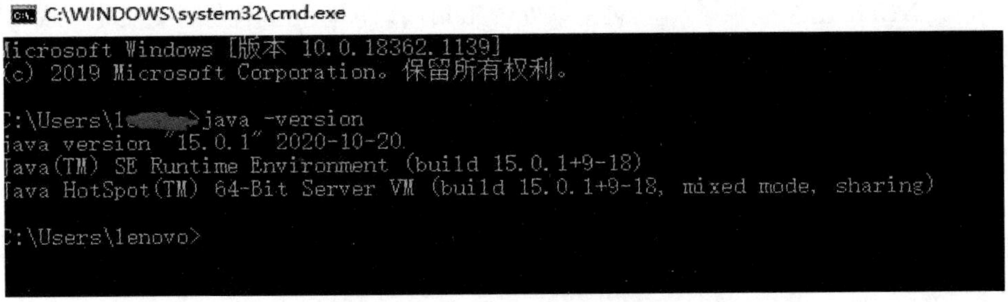

图 5-10　查看 JAVA 程序版本信息界面

（2）在 DOS 提示符下输入"Pip install stanfordcorenlp"后按"Enter"键，安装 Stanford CoreNLP 程序。

（3）下载 Stanford CoreNLP 文件。下载界面如图 5-11 所示。

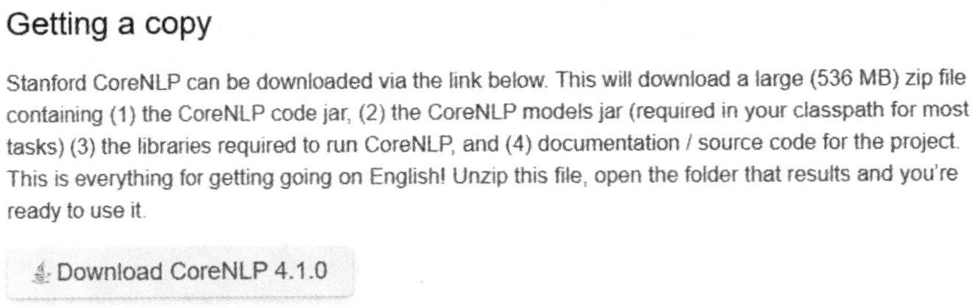

图 5-11　Stanford CoreNLP 文件下载界面

（4）下载中文/英文模型 jar 包，下载界面如图 5-12 所示。

（5）将第(3)步中下载的 zip 文件解压，然后将第(4)步中下载的中文、英文 jar 包放到该文件夹中，界面如图 5-13 所示。

将原汉语模型文件名改为 stanford-chinese-corenlp-2020-11-04-models，时间可以自己定，英文模型文件命名方法类似。

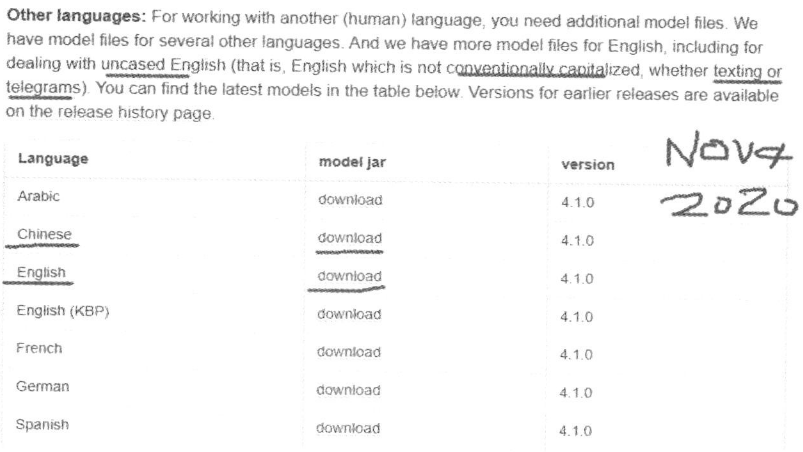

Other languages: For working with another (human) language, you need additional model files. We have model files for several other languages. And we have more model files for English, including for dealing with uncased English (that is, English which is not conventionally capitalized, whether texting or telegrams). You can find the latest models in the table below. Versions for earlier releases are available on the release history page.

Language	model jar	version
Arabic	download	4.1.0
Chinese	download	4.1.0
English	download	4.1.0
English (KBP)	download	4.1.0
French	download	4.1.0
German	download	4.1.0
Spanish	download	4.1.0

Nov 4 2020

图 5-12　Stanfordcorenlp 中文/英文模型 jar 包下载界面

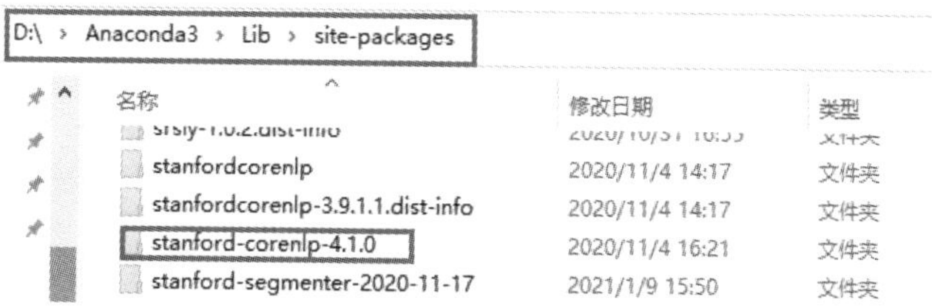

图 5-13　配置 Stanfordcorenlp 模型文件界面

如果 Python 是用 Anaconda 安装的，把第(3)步的文件夹(stanford-corenlp-4.1.0)放在 D:\Anaconda3\Lib\site-packages 文件夹内(以 Anaconda 安装在 D 盘目录下为例)。

(6) 测试 Stanford CoreNLP 安装与配置，在 Python 编译器里输入如图 5-14 界面所示的代码。

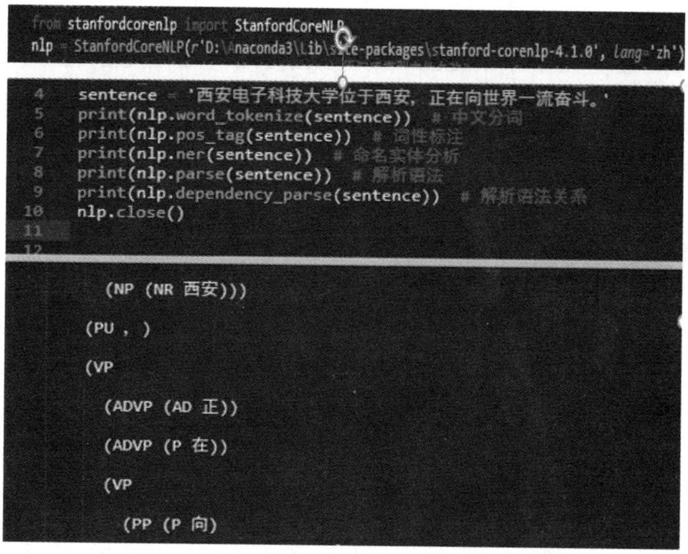

图 5-14　测试 Stanfordcorenlp 安装与配置界面

(7) 报错问题。在第(5)步中未改文件名会导致报错。如果程序提示以下报错信息：

Traceback (most recent call last):
　File "stanford_visualize.py", line 4, in <module>

```
        with StanfordCoreNLP(r'stanford-corenlp-4.1.0', lang='zh') as nlp:
    File "/home/appleyuchi/anaconda3/envs/Python3.6/lib/python3.6/site-packages/stanfordcorenlp/corenlp.py", line 74, in __init__
        self.lang) + ' not exists. You should download and place it in the ' + directory + ' first.')
OSError: stanford-chinese-corenlp-yyyy-MM-dd-models.jar not exists.
You should download and place it in the stanford-corenlp-4.1.0/ first.
```

解决方法如下：

将第(4)步中下载的中、英文模型文件 stanford-corenlp-4.1.0-models-chinese.jar 和 stanford-corenlp-4.1.0-models-english.jar 分别按照以下格式命名为：

stanford-chinese-corenlp-2020-11-05-models.jar

stanford-english-corenlp-2020-11-05-models.jar

重新运行程序，报错就会消失。

(8) 开启 CoreNLP 的 API 服务，界面如图 5-15 所示。

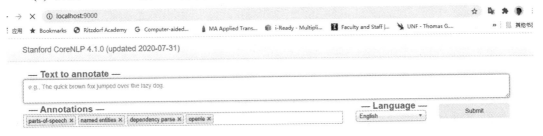

图 5-15　开启 CoreNLP 的 API 服务界面

2. Stanford CoreNLP 短语结构文法句法分析

利用 Standford CoreNLP，使用短语结构文法分析汉语、英语句法。

(1) StanfordCoreNLP 短语结构文法分析汉语句法。示例代码如下：

```
from stanfordcorenlp import StanfordCoreNLP
zh_nlp = StanfordCoreNLP(r'D:\Anaconda3\Lib\site-packages\stanford-corenlp-4.1.0', lang='zh')
sent_ZH = '西安电子科技大学位于西安，正在向世界一流奋斗。'
## 4  句法成分分析(Constituency Parse)
print ('Constituency Parsing:', zh_nlp.parse(sent_ZH) + "\n")
zh_nlp.close()
```

输出：

```
Constituency Parsing: (ROOT
    (IP
      (IP
        (NP
          (NP (NR 西安))
          (NP (NN 电子) (NN 科技) (NN 大学)))
        (VP
          (VP (VV 位于)
```

```
            (NP (NR 西安)))
          (PU ，)
          (VP
            (ADVP (AD 正))
            (ADVP (P 在))
            (VP
              (PP (P 向)
                (NP
                  (NP (NN 世界))
                  (ADJP (JJ 一流))
                  (NP (NN 奋斗))))))))
        (PU 。)))
```

(2) Stanford CoreNLP 短语结构文法分析英语句法。示例代码如下：

```
from stanfordcorenlp import StanfordCoreNLP
en_nlp = StanfordCoreNLP(r'D:\Anaconda3\Lib\site-packages\stanford-corenlp-4.1.0', lang='en')
sent_EN = "If you want to change the source code and recompile the files, see these instructions"
print ('Constituency Parsing:', en_nlp.parse(sent_EN))
en_nlp.close()
```

输出：

```
Constituency Parsing: (ROOT
  (NP
    (S
      (SBAR (IN If)
        (S
          (NP (PRP you))
          (VP (VBP want)
            (S
              (VP (TO to)
                (VP
                  (VP (VB change)
                    (NP (DT the) (NN source) (NN code)))
                  (CC and)
                  (VP (VB recompile)
                    (NP (DT the) (NNS files))))))))))
      (, ,)
      (VP (VB see)
        (NP (DT these) (NNS instructions))))))
```

3. Stanford CoreNLP 依存文法句法分析

利用 StandfordCoreNLP，使用依存文法分析汉语、英语句法。示例代码如下：

(1) StanfordCoreNLP 依存文法分析汉语句法。示例代码如下：

```
from stanfordcorenlp import StanfordCoreNLP
zh_nlp = StanfordCoreNLP(r'D:\Anaconda3\Lib\site-packages\stanford-corenlp-4.1.0', lang='zh')
sent_ZH = '西安电子科技大学位于西安，正在向世界一流奋斗。'
print ('Dependency:', zh_nlp.dependency_parse(sent_ZH))
zh_nlp.close()
```

输出：Dependency: [('ROOT', 0, 5), ('compound:nn', 4, 1), ('compound:nn', 4, 2), ('compound:nn', 4, 3), ('nsubj', 5, 4), ('dobj', 5, 6), ('punct', 5, 7), ('conj', 5, 8), ('ccomp', 8, 9), ('case', 13, 10), ('compound:nn', 13, 11), ('amod', 13, 12), ('nmod:prep', 9, 13), ('punct', 5, 14)]

(2) StanfordCoreNLP 依存文法分析英语句法。示例代码如下：

```
from stanfordcorenlp import StanfordCoreNLP
en_nlp = StanfordCoreNLP(r'D:\Anaconda3\Lib\site-packages\stanford-corenlp-4.1.0', lang='en')
sent_EN = "If you want to change the source code and recompile the files, see these instructions"
print ('Dependency:', en_nlp.dependency_parse(sent_EN))
en_nlp.close()
```

输出：

Dependency: [('ROOT', 0, 14), ('mark', 3, 1), ('nsubj', 3, 2), ('advcl', 14, 3), ('mark', 5, 4), ('xcomp', 3, 5), ('det', 8, 6), ('compound', 8, 7), ('obj', 5, 8), ('cc', 10, 9), ('conj', 5, 10), ('det', 12, 11), ('obj', 10, 12), ('punct', 14, 13), ('det', 16, 15), ('obj', 14, 16)]

部分依存文法符号的解释如表 5-1 所示。

表 5-1 部分依存文法符号及其描述

符 号	描 述
ROOT	句子的核心，整个句法结构的根节点
IP	简单从句
NP	名词短语
VP	动词短语
PU	断句符，包括句号、问号、感叹号等标点符号
LCP	方位词短语
PP	介词短语
CP	从句结构
DNP	限定词组结构
ADVP	副词短语
ADJP	形容词短语
DP	限定词短语
QP	量词短语
NN	常用名词
NR	固有名词

5.3 可视化句法分析

(1) 使用 NLTK 和 StanfordCoreNLP 分析汉语并可视化。代码如下：

```
from nltk.tree import Tree
from stanfordcorenlp import StanfordCoreNLP
test_vis_sent = '我是来测试句法分析可视化的'
with StanfordCoreNLP(r'D:\Anaconda3\Lib\site-packages\stanford-corenlp-4.1.0', lang='zh') as zh_nlp:
    Tree.fromstring(zh_nlp.parse(test_vis_sent)).draw()
```

输出树图结果如图 5-16 所示。

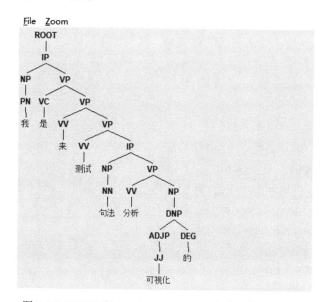

图 5-16　NLTK 和 StanfordCoreNLP 可视化分析汉语结果

(2) 输入要分析的句子，选择依存文法解析和语言后提交。CoreNLP 的 API 句法分析界面如图 5-17 所示。

图 5-17　CoreNLP 的 API 句法分析界面

句法分析结果如图 5-18 所示。

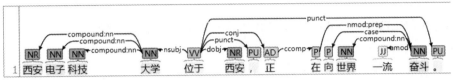

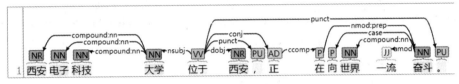

图 5-18　CoreNLP 的 API 句法分析结果图

(3) 使用 spaCy 来显示英语句法依存关系。代码如下：

```
import spacy
from spacy import displacy


nlp = spacy.load("en_core_web_sm")
doc = nlp( "The cat and dog are cheerfully chasing each other." )
svg = displacy.render(doc, style="dep")
# output_path = Path('filepath\\cat_dog_sent.svg')#保存图示在指定路径文件夹内
output_path = Path("./cat_dog_sent_plot.svg")#保存图示在程序所在文件夹内
output_path.open("w", encoding="utf-8").write(svg)
displacy.serve(doc, style='dep')
```

依存结构图如图 5-19 所示。

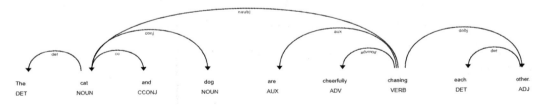

图 5-19　spaCy 句法分析依存结构图

第 6 章　正则表达式

6.1　正则表达式的概念、用法及用途

1. 正则表达式的概念

正则表达式(Regular Expression，在代码或软件中常简写为 regex、regexp 或 RE)是计算机科学中的一个概念。正则表达式是操作字符串的一种逻辑公式，它用事先定义好的一些特定字符或字符组合组成"规则字符串"来表达对字符串的一种过滤逻辑。

正则表达式是记录文本规则的代码，可描述字符模式的对象，用来匹配、检索和替换字符串模式。正则表达式是操作字符串的强大工具，在处理文本和提取网页数据时常常会用到，许多软件和程序设计语言都支持利用正则表达式进行字符串操作。

2. 正则表达式的用法及用途

日常使用的 NotePad、EditPad Pro、PowerGrep 等工具都支持正则表达式查找、替换等功能。以在 NotePad 中使用正则表达式为例，用 NotePad 打开文本，使用快捷键 Ctrl + H 查找替换，显示界面如图 6-1 所示。

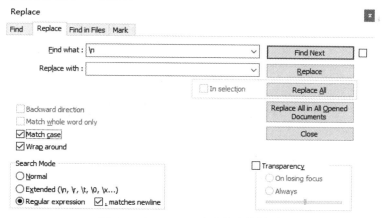

图 6-1　在 NotePad 中查找替换界面

勾选 Regular expression 即可使用正则表达式进行搜索、替换操作。图中"\n"是搜索换行符。

其他常见正则表达式特殊字符的含义如下：

(1) .：匹配除"\n"之外的任何单个字符。要匹配包括"\n"在内的任何字符，可使用"[.\n]"的模式。

(2) *：匹配 0 个或多个表达式。

(3) ?：匹配 0 个或 1 个由前面的正则表达式定义的片段。

(4) \：转义符，和其他字符配合使用，把特殊符号变成普通符号，把普通符号变成特殊符号。

(5) ()：提取括号里面的内容。

(6) \d：匹配一个数字字符，等价于[0-9]。

(7) \D：匹配一个非数字字符，等价于[^0-9]。

(8) \s：匹配任何空白字符，包括空格、制表符、换页符等，等价于[\f\n\r\t\v]。

(9) \S：匹配任何非空白字符，等价于[^\f\n\r\t\v]。

(10) \w：匹配包括下画线的任何单词字符，等价于[A-Za-z0-9_]。

(11) \W：匹配任何非单词字符，等价于[^A-Za-z0-9_]。

下面以表示单独列出一组字符[...]匹配字符为例，如表 6-1 所示。

表 6-1 正则表达式匹配字符示例

实例	描述
notepad	匹配"notepad"
[Nn]otepad	匹配"Notepad"或"notepad"
ro[cd]	匹配"roc"或"rod"
[cfdeabc]	匹配中括号内的任意一个字母
[0-9]	匹配任何数字，类似于[0123456789]
[a-z]	匹配任何小写字母
[A-Z]	匹配任何大写字母
[a-zA-Z0-9]	匹配任何字母及数字
[^defabcd]	匹配除了 defabcd 字母以外的所有字符
[^0-9]	匹配除了数字外的字符

6.2 Python 中使用正则表达式

1. 查询正则表达式函数

导入 re 库，查询正则表达式函数，代码如下：

```
import re
print(dir(re))
输出：['A', 'ASCII', 'DEBUG', 'DOTALL', 'I', 'IGNORECASE', 'L', 'LOCALE', 'M', 'MULTILINE', 'Match', 'Pattern', 'RegexFlag', 'S', 'Scanner', 'T', 'TEMPLATE', 'U', 'UNICODE', 'VERBOSE', 'X', '_MAXCACHE', '__all__', '__builtins__', '__cached__', '__doc__', '__file__', '__loader__', '__name__', '__package__', '__spec__', '__version__', '_cache', '_compile', '_compile_repl', '_expand', '_locale', '_pickle', '_special_chars_map', '_subx', 'compile', 'copyreg', 'enum', 'error', 'escape', 'findall', 'finditer', 'fullmatch', 'functools', 'match', 'purge', 'search', 'split', 'sre_compile', 'sre_parse', 'sub', 'subn', 'template']
```

使用正则表达式前需要寻找字符串规律，之后用正则表达式表示该规律，便可操作符合该规律的字符串。例如，XML 网页文件标题的首尾部分都有<title>字符串，用(.*?)表示标题内容，使用正则表达式"<title>(.*?)</title>"就可以将网页的标题提取出来。

2. re 模块函数的语法

re 模块函数的语法格式说明如表 6-2 所示。

表 6-2　re 模块函数的语法格式说明

re 模块函数的语法格式	说　　明
compile(pattern, flag=0)	编译正则表达式 pattern，然后返回一个正则表达式对象
match(pattern, string, flags=0)	使用带有可选标记的正则表达式模式来匹配字符串。匹配成功，返回匹配对象；匹配失败，返回 None
search(pattern, string, flags=0)	使用可选标记搜索字符串中第一次出现的正则表达式模式。匹配成功，则返回匹配对象；匹配失败，则返回 None
findall(pattern, string, [, flags])	查找字符串中出现的所有正则表达式，并返回一个列表
split(pattern, string, max=0)	根据正则表达式的模式分割，spilt 函数将字符串分割为列表，然后返回成功匹配的列表，分割最多操作 max 次(默认分割所有匹配成功的位置)
sub(pattern, repl, string, count=0)	使用 repl 替换所有正则表达式的模式在字符串中出现的位置，除非定义 count，否则将替换所有出现的位置

3. re 函数的使用

(1) re.compile 函数。compile 函数用于编译正则表达式，生成一个正则表达式(Pattern)对象，供 match()和 search()这两个函数使用。语法格式如下：

re.compile(pattern[, flags])

其中，pattern 为一个字符串形式的正则表达式。flags 可选，表示匹配模式，比如忽略大小写，多行模式等，具体参数含义如下：

① re.I：忽略大小写。
② re.L：表示特殊字符集 \w、\W、\b、\B、\s、\S 依赖当前环境。
③ re.M：多行模式匹配，影响 ^ 和 $。
④ re.S：使 "." 匹配包括换行在内的所有字符。
⑤ re.U：根据 Unicode 字符集解析字符，表示特殊字符集 \w、\W、\b、\B、\d、\D、\s、\S。
⑥ re.X：为了增加可读性，忽略空格和 # 后面的注释。示例代码如下：

```
import re
str1 = "12https://c.runoob.com/front-end/854StanfordCoreNLP 依存句法分析(Dependency Parse) masxx@yahoo.com,18881882123"
pattern = re.compile(r'\d+') # 用于匹配至少一个数字
digt = pattern.match(str1)
print(digt)
```

输出：
    <re.Match object; span=(0, 2), match='12'>

digt1 = pattern.match(str1, 2, 10) # 从 'h' 的位置开始匹配，没有匹配

print(digt1)

输出：
    None

pattern = re.compile('([0-9]+)')

digt2 = pattern.match(str1)

print(digt2)

输出：
    <re.Match object; span=(0, 2), match='12'>

(2) re.search 函数。re.search 函数用于扫描整个字符串并返回第一个成功的匹配。示例代码如下：

digt3 = re.search(r'\d+', str1)

print(digt3)

输出：
    <re.Match object; span=(0, 2), match='12'>

print(digt3.group(0))

输出：
    12

digt4 = pattern.search(str1)

print(digt4)

输出：
    <re.Match object; span=(0, 2), match='12'>

str2 = "12https://c.runoob.com/front-end/854StanfordCoreNLP 依存句法分析(Dependency Parse)masxx@yahoo.com,18881882123"

reslt = re.search('Parse\)(.*?)\,188', str2)

print(reslt)

输出：
    <re.Match object; span=(70, 95), match='Parse)masxx@yahoo.com,188'>

print(reslt.group(0))

输出：
    Parse)masxx@yahoo.com,188

print(reslt.group(1))

输出：
    masxx@yahoo.com

print(reslt.group())

输出：
　　Parse)masxx@yahoo.com,188

(3) 同时匹配字符串中的多个值。示例代码如下：
str3 = '姓名-王汪，wechat:11118876 blog:https://www.abcc.com/wangwww/ 欢迎联系'
res = re.search(r'wechat:(\d+) blog:(.*?) 欢迎联系', str3)
print(res)
输出：
　　<re.Match object; span=(6, 61), match='wechat:11118876 blog:https://www.abcc.com/wangwww>
同时取出三个值，返回的是元组()
result = res.group(0, 1, 2) if res else None
print(result)
输出：
　　('wechat:11118876 blog:https://www.abcc.com/wangwww/ 欢迎联系', '11118876', 'https://www.abcc.com/wangwww/')

(4) group()可以一次输入多个组号，它将返回一个包含那些组所对应值的元组。示例代码如下：
res1, res2, res3 = res.group(0, 1, 2) if res else None
print(res1)
输出：wechat:11118876 blog:https://www.abcc.com/wangwww/ 欢迎联系
print(res2)
输出：
　　11118876
print(res3)
输出：
　　https://www.abcc.com/wangwww/

group()和group(0)一样，均可获取匹配的整个字符串，group(1)取出匹配的第一个值，group(2)取出匹配的第二个值；group()与group(0, 1, 2)一样，同时取出上面对应的三个值，返回元祖。

(5) re.findall 返回字符串中所有不重叠匹配项的列表。示例代码如下：
num = re.findall('\d+',str1)
print(num)
输出：
　　['12', '854', '18881882123']

findall()返回的是括号所匹配到的结果(如patt1)，多个括号就会返回多个括号分别匹配到的结果(如patt)，如果没有括号就返回整条语句所匹配到的结果(如patt2)。示例代码如下：
import re

strng="dfg　bge　acf　agf"

```
patt=re.compile("((\w+)\s+\w+)")
print(patt.findall(strng))
```
输出:

  [('dfg bge', 'dfg'), ('acf agf', 'acf')]

```
patt1=re.compile("(\w+)\s+\w+")
print(patt1.findall(strng))
```
输出:

  ['dfg', 'acf']

```
patt2=re.compile("\w+\s+\w+")
print(patt2.findall(strng))
```
输出:

  ['dfg bge', 'acf agf']

(6) ".", "*", "?" 及其组合用法。

```
import re

strn = 'abccadabbcdefaeebacadaabbc'
```

① "." 匹配除 \n(换行符)以外的任意一个字符。示例代码如下:

```
print(re.findall('a.b',strn))
```
输出:

  ['abb', 'aab']

② "*" 匹配表达式前面出现 0 次或任意次的字符。示例代码如下:

```
print(re.findall('a*b',strn))
```
输出:

  ['ab', 'ab', 'b', 'b', 'aab', 'b']

③ ".*" 匹配表达式之前字符串与表达式之后字符串之间的所有内容。示例代码如下:

```
print(re.findall(r'a.*b',strn))
```
输出:

  ['abccadabbcdefaeebacadaabb']

④ ".*?" 遇到开始和结束就进行截取,截取多次符合的结果,中间没有字符也会被截取。示例代码如下:

```
print(re.findall('a.*?b',strn))
```
输出:

  ['ab', 'adab', 'aeeb', 'acadaab']

⑤ "(.*?)" 只保留括号里的内容。示例代码如下:

```
print(re.findall('a(.*?)b',strn))
```
输出:

  ['', 'da', 'ee', 'cadaa']

(7) Re.S 的用法。示例如下:

```
strn1 = '''aabbab
        aabbaa
        bb'''
```

只搜索换行符之前的字符串,遇到换行符停止搜索。

print(re.findall('a.*?b',strn1))

输出:

['aab', 'ab', 'aab']

re.S 不受换行符 \n 的影响,继续搜索换行符后的内容。

print(re.findall('a.*?b',strn1,re.S))

输出:

['aab', 'ab', 'aab', 'aa\n   b']

(8) re.split 函数的用法。示例代码如下:

```
import re
s = '''abc 22 dee 66
fff 221 ggg
hhh 333 hhh, www, com, http; router; hub'''
```

① 按照 ","或";"切分。示例代码如下:

print(re.split(r'\,|\;',s))

输出:

['abc 22 dee 66\nfff 221 ggg\nhhh 333 hhh', ' www', ' com', ' http', ' router', ' hub']

② 按照 ',' ,';' 或数字切分。示例代码如下:

print(re.split(r'\,|\d|\;',s))

输出:

['abc ', '', ' dee ', '', '\nfff ', '', '', ' ggg\nhhh ', '', '', ' hhh', ' www', ' com', ' http', ' router', ' hub']

③ 按照数字切分。示例代码如下:

print(re.split(r'\d+', s))

输出:

['abc ', ' dee ', '\nfff ', ' ggg\nhhh ', ' hhh, www, com, http; router; hub']

④ \.+ 匹配不到,返回包含自身的列表。示例代码如下:

print(re.split(r'\.+', s, 1))

输出:

['abc 22 dee 66\nfff 221 ggg\nhhh 333 hhh, www, com, http; router; hub']

⑤ maxsplit 参数用法。示例代码如下:

print(re.split(r'\d+', s, 1))

输出:

['abc ', ' dee 66\nfff 221 ggg\nhhh 333 hhh, www, com, http; router; hub']

(9) re.sub 函数的用法。示例代码如下:

sub(pattern, repl, string, count=0, flags=0)

替换函数是指将正则表达式 pattern 匹配到的字符串替换为 repl 指定的字符串，参数 count 用于指定最大替换次数。其中三个必选参数为 pattern、repl、string，两个可选参数为 count、flags。

① 将目标字符串替换为指定字符串。示例代码如下：

```
import re
s = "the sum of 7 and 9 is [7+9]."
print(re.sub(r'\[7\+9\]', '16', s))
```

输出：
  the sum of 7 and 9 is 16.

② 用正则表达式替换字符串中所有数字。示例代码如下：

```
S1 = "two numbers are 112 and 333 in the set."
ReStr = re.sub(r'\d+', 'unknown numbers', S1)
print(ReStr)
```

输出：
  two numbers are unknown numbers and unknown numbers in the set.

③ 参数 pattern 表示正则表达式中的模式字符串。反斜杠加数字(\N)，对应着匹配的组(matched group)，如 \3 表示匹配前面 pattern 中的第 3 个 group。

例如，要处理"hi Fansy, oh Fansy"字符串中的 Fansy，若要把整个字符串换成 Osaka。代码如下：

```
Import re
inpStr = "hi Fansy, oh Fansy"
redStr = re.sub(r"hi (\w+), oh \1", "Osaka", inpStr)
print("replacedStr=",redStr)
```

输出：
  replacedStr= Osaka

代码中的 \1 指的是(\w+)，也就是任何单词字符。

④ 参数 repl 是被替的字符串，可以是字符串或函数。若 repl 是字符串的话，其中任何反斜杠转义字符都会被处理。例如，\n 会被处理为对应的换行符；\r 会被处理为回车符；其他不能识别的转移字符，则只是被识别为普通的字符。

例如，\h 会被处理为 h 这个字母本身；反斜杠加 g 以及尖括号内一个名字，即 \g<name> 对应已命名的组。代码如下：

```
inpStr = "hi Fansy, oh Fansy"
redStr1 = re.sub(r"hi (\w+), oh \1", "\g<1>", inpStr)
print("replacedStr1=",redStr1)
```

输出：
  replacedStr1= Fansy

对应的带命名的组(named group)的版本是：

```
inpStr = "hi Fansy, oh Fansy"
redStr2 = re.sub(r"hi (?P<name>\w+), oh (?P=name)", "\g<name>", inpStr
```

```
print("replacedStr=",redStr2)
```
输出：
```
replacedStr= Fansy
```
⑤ 参数 count 指替换次数，下面的句子中有 5 个"Fansy"，若只替换前两个，则指定 count 值为 2。示例代码如下：
```
inpStr = "hi Fansy, oh Fansy,ohh Fansy, gosh Fansy, OMG Fansy"
redStr3 = re.sub(r"Fansy", "Osaka", inpStr,2)
print("replacedStr=",redStr3)
```
输出：
```
replacedStr= hi Osaka, oh Osaka,ohh Fansy, gosh Fansy, OMG Fansy
```

6.3 使用正则表达式爬取网页数据

爬取网页数据的基本方法是模仿浏览器爬取网页数据并使用正则表达式提取需要的文本、图片或视频等，或者使用代理模仿浏览器爬取数据后再提取所需信息。下面通过两个例子来说明其具体步骤。

【例 6-1】 模仿浏览器爬取网页数据并提取文本。代码如下：

```python
# 导入正则表达式和请求库
import re
import requests

# 导入 PyQuery 用于解析 DOM 结构
from pyquery import PyQuery as pq

# 设置 User-Agent 请求头，模拟浏览器
headers = {    "User-Agent": "Mozilla/5.0 (Windows NT 10.0; WOW64) AppleWebKit/537.36 (KHTML, like Gecko) Chrome/67.0.3396.99 Safari/537.36",
Accept": "text/html,application/xhtml+xml,application/xml;q=0.9,image/webp,*/*;q=0.8",    "Accept-Language": "zh-CN,zh;q=0.9,en;q=0.8",    "Connection": "keep-alive"}

# 目标网页 URL
web_url = 'https://www.163.com/tech/article/JF5V9FJG00097U7T.html'

# 获取网页内容
data = requests.get(web_url,headers=headers)
data.encoding = 'utf-8'
resove_web_url = data.text
```

```python
# 根据网页结构，构建正则表达式匹配需要内容
pat = '<div class="post_body">(.*?)</div>'

# 找到匹配内容
rslt = re.compile(pat, re.S).findall(data.text)

# 将列表转换为字符串
string = rslt[0]

# 用 PyQuery 删除 HTML 标签
text2 = pq(string).text()

# 将内容写入文本文件
fileOb = open('news.txt','w',encoding='utf-8')
fileOb.write(text2)
fileOb.close()
```

部分保存文件内容：

10月23日消息，美国时间周二，由前 OpenAI 研究高管创立并获得亚马逊支持的人工智能初创企业 Anthropic 宣布，该公司已在人工智能领域取得了一个里程碑式的突破，其研发的 AI 智能体已具备与人类相当的能力，可以操作电脑完成更复杂的任务。

……

【例6-2】 使用代理模仿浏览器爬取网页数据并提取图片链接，代码如下：

```python
# 导入正则表达式、请求、urllib 库
import re
import requests
import urllib.request

# 设置 User-Agent 请求头
headers={"UserAgent":"Mozilla/5.0 (Windows NT 10.0; WOW64) AppleWebKit/537.36 (KHTML, like Gecko) Chrome/67.0.3396.99 Safari/537.36"}

# 设置代理 IP
proxies = {"http":"http://127.0.0.1:54182","https":"https://127.0.0.1:53935"}

# 循环抓取多个页面
for i in range(0,3):

    # 构造目标网页 URL
    web_url = 'http://search.jumei.com/?filter=0-11-"+str(i)+"&search=%E5%8F%A3%E7%BA%A2&bid=
```

4&site=bj'

# 获取网页内容
data = requests.get(web_url,headers=headers, proxies=proxies)
data.encoding = 'utf-8'
resove_web_url = data.text

# 构建正则表达式，提取图片链接
pat = 'original="(.*?)"'
rslt = re.compile(pat, re.S).findall(data.text)

# 打印并统计结果
print(rslt)
输出：
　　['http://p3.jmstatic.com/product/005/223/5223038_std/5223038_1_60_60.jpg',
　　'http://p3.jmstatic.com/product/005/223/5223038_std/5223038_2_60_60.jpg',
　　'http://p3.jmstatic.com/product/005/223/5223038_std/5223038_3_60_60.jpg',
　　'http://p3.jmstatic.com/product/005/223/5223038_std/5223038_4_60_60.jpg',
　　…
　　'http://p1.jmstatic.com/product/004/897/4897816_std/4897816_350_350.jpg']
print(len(rslt))
输出：
　　177

6.4 常用的正则表达式

1. 校验数字的表达式

(1) 数字：^[0-9]*$。
(2) n 位的数字：^\d{n}$。
(3) 至少 n 位的数字：^\d{n,}$。
(4) m～n 位的数字：^\d{m,n}$。
(5) 零和非零开头的数字：^(0|[1-9][0-9]*)$。
(6) 非零开头的最多带两位小数的数字：^([1-9][0-9]*)+(\.[0-9]{1,2})?$。
(7) 带 1～2 位小数的正数或负数：^(\-)?\d+(\.\d{1,2})$。
(8) 正数、负数和小数：^(\-|\+)?\d+(\.\d+)?$。
(9) 有两位小数的正实数：^[0-9]+(\.[0-9]{2})?$。
(10) 有 1～3 位小数的正实数：^[0-9]+(\.[0-9]{1,3})?$。
(11) 非零的正整数：^[1-9]\d*$ 或 ^([1-9][0-9]*){1,3}$ 或 ^\+?[1-9][0-9]*$。
(12) 非零的负整数：^\-[1-9][]0-9"*$ 或 ^-[1-9]\d*$。

(13) 非负整数：^\d+$ 或 ^[1-9]\d*|0$。

(14) 非正整数：^-[1-9]\d*|0$ 或 ^((-\d+)|(0+))$。

(15) 非负浮点数：^\d+(\.\d+)?$ 或 ^[1-9]\d*\.\d*|0\.\d*[1-9]\d*|0?\.0+|0$。

(16) 非正浮点数：^((-\d+(\.\d+)?)|(0+(\.0+)?))$ 或 ^(-([1-9]\d*\.\d*|0\.\d*[1-9]\d*))|0?\.0+|0$。

(17) 正浮点数：^[1-9]\d*\.\d*|0\.\d*[1-9]\d*$ 或 ^(([0-9]+\.[0-9]*[1-9][0-9]*)|([0-9]*[1-9][0-9]*\.[0-9]+)|([0-9]*[1-9][0-9]*))$。

(18) 负浮点数：^-([1-9]\d*\.\d*|0\.\d*[1-9]\d*)$ 或 ^(-(([0-9]+\.[0-9]*[1-9][0-9]*)|([0-9]*[1-9][0-9]*\.[0-9]+)|([0-9]*[1-9][0-9]*)))$。

(19) 浮点数：^(-?\d+)(\.\d+)?$ 或 ^-?([1-9]\d*\.\d*|0\.\d*[1-9]\d*|0?\.0+|0)$。

2. 校验字符的表达式

(1) 汉字：^[\u4e00-\u9fa5]{0,}$。

(2) 英文和数字：^[A-Za-z0-9]+$ 或 ^[A-Za-z0-9]{4,40}$。

(3) 长度为3～20的所有字符：^.{3,20}$。

(4) 由26个英文字母组成的字符串：^[A-Za-z]+$。

(5) 由26个大写英文字母组成的字符串：^[A-Z]+$。

(6) 由26个小写英文字母组成的字符串：^[a-z]+$。

(7) 由数字和26个英文字母组成的字符串：^[A-Za-z0-9]+$。

(8) 由数字、26个英文字母或者下画线组成的字符串：^\w+$ 或 ^\w{3,20}$。

(9) 包含中文、英文、数字、下画线：^[\u4E00-\u9FA5A-Za-z0-9_]+$。

(10) 中文、英文、数字，但不包含下画线等符号：^[\u4E00-\u9FA5A-Za-z0-9]+$ 或 ^[\u4E00-\u9FA5A-Za-z0-9]{2,20}$。

(11) 可以输入含有 ^%&',;=?$\" 等字符：[^%&',;=?$\x22]+。

(12) 禁止输入含有~的字符：[^~\x22]+。

3. 特殊需求表达式

(1) Email 地址：^\w+([-+.]\w+)*@\w+([-.]\w+)*\.\w+([-.]\w+)*$。

(2) 域名：[a-zA-Z0-9][-a-zA-Z0-9]{0,62}(\.[a-zA-Z0-9][-a-zA-Z0-9]{0,62})+\.?。

(3) 互联网 URL：[a-zA-z]+://[^\s]* 或 http://([\w-]+\.)+[\w-]+(/[\w-./?%&=]*)?$。

(4) 手机号码：^(13[0-9]|14[5|7]|15[0|1|2|3|4|5|6|7|8|9]|18[0|1|2|3|5|6|7|8|9])\d{8}$。

(5) 电话号码("XXX-XXXXXXX"、"XXXX-XXXXXXXX"、"XXX-XXXXXXXX"、"XXX-XXXXXXXX"、"XXXXXXX" 和 "XXXXXXXX")：^(\(\d{3,4}-)|\d{3.4}-)?\d{7,8}$。

(6) 国内电话号码(0522-22256567、029-86929988)：\d{3}-\d{8}|\d{4}-\d{7}。

(7) 电话号码正则表达式(支持手机号码3～4位区号、7～8位直播号码、1～4位分机号)：((\d{11})|^((\d{7,8})|(\d{4}|\d{3})-(\d{7,8})|(\d{4}|\d{3})-(\d{7,8})-(\d{4}|\d{3}|\d{2}|\d{1})|(\d{7,8})-(\d{4}|\d{3}|\d{2}|\d{1}))$)。

(8) 身份证号(15或18位数字)，最后一位是校验位，可能为数字或字符X：(^\d{15}$)|(^\d{18}$)|(^\d{17}(\d|X|x)$)。

(9) 账号(字母开头，长度为5～16字节，允许字母数字下画线)：^[a-zA-Z][a-zA-Z0-9_]{4,15}$。

(10) 密码(以字母开头,长度为6～18字符,只能包含字母、数字和下画线):^[a-zA-Z]\w{5,17}$。

(11) 强密码(包含大小写字母和数字的组合,不能使用特殊字符,长度为8～10字符):^(?=.*\d)(?=.*[a-z])(?=.*[A-Z])[a-zA-Z0-9]{8,10}$。

(12) 强密码(包含大小写字母和数字的组合,可以使用特殊字符,长度在8～10字符):^(?=.*\d)(?=.*[a-z])(?=.*[A-Z])(?=.*[@#$%^&+=])(?=.{8,10}$).*$。

(13) 日期格式:^\d{4}-\d{1,2}-\d{1,2}。

(14) 一年的12个月(01～09和1～12):^(0?[1-9]|1[0-2])$。

(15) 一个月的31天(01～09和1～31):^((0?[1-9])|((1|2)[0-9])|30|31)$。

(16) 金额的正则表达式:((?<!\d)\d([ ]*\d([ ]*\d))?(?:[ ]*,?([ ]*\d))*(\.[ ]*\d[ ]*\d(?![ ]*\d))?|(\.[ ]*\d[ ]*\d(?![ ]*\d)))。

(17) xml文件:^([a-zA-Z]+-?)+[a-zA-Z0-9]+\\.[x|X][m|M][l|L]$。

(18) 中文字符的正则表达式:[\u4e00-\u9fa5]。

(19) 双字节字符:[^\x00-\xff](包括汉字在内,可以用来计算字符串的长度(一个双字节字符长度计2,ASCII字符计1))。

(20) 空白行的正则表达式:\n\s*\r(可以用来删除空白行)。

(21) HTML标记的正则表达式:<(\S*?)[^>]*>.*?|<.*?/>。

(22) 首尾空白字符的正则表达式:^\s*|\s*$ 或 (^\s*)|(\s*$)(可以用来删除行首、行尾的空白字符(包括空格、制表符、换页符等)。

(23) 腾讯QQ号:[1-9][0-9]{4,}。

(24) 中国邮政编码:[1-9]\d{5}(?!\d)。

(25) IPv4地址:((2(5[0-5]|[0-4]\d))|[0-1]?\d{1,2})(\.((2(5[0-5]|[0-4]\d))|[0-1]?\d{1,2})){3}。

注意: ^ 和 $ 分别为正则表达式的开始和结束边界。

第 7 章 文 本 处 理

7.1 提取中、英文

使用正则表达式提取文本中的中、英文。代码示例如下:

```
import re
fob = open('new-file1.txt','r',encoding="utf-8")


for line in fob.readlines():
    line = line.strip()
    if line == '\n':
        line = line.strip("\n")


    pattern = "[^\u4e00-\u9fa5]"    # 提取结果为英语、数字及符号
    regex = re.compile(pattern)
    rlt =   regex.findall(line)
        rlt1 = []
    for i in rlt:
    rlt1.append(i)
    rlt2 = ''.join(rlt1)
    print(rlt2.encode('gbk','ignore').decode('gbk'))
```

运行以上代码将会提取出文本中的英文、数字及符号。如果需要提取文本中的汉语,在以上代码中使用 pattern = "[\u4e00-\u9fa5]" 即可提取。

7.2 提取词汇和短语

提取词汇和短语的常见工具有 NLTK、TextBlob 和 spaCy 等。这三种工具的特点不尽相同,下面逐个介绍它们的使用方法。

1. NLTK 提取名词和名词短语

1) NLTK 提取名词

在使用 NLTK 提取名词时可以利用词性标签,通过遍历的方式提取,示例代码如下:

```
import nltk

File = open(r'CHNC-clean-para500.txt',encoding = "utf-8") #open file
lines = File.read() #read all lines
sentences = nltk.sent_tokenize(lines) #tokenize sentences
nouns = [] #empty to array to hold all nouns

for sentence in sentences:
    for word,pos in nltk.pos_tag(nltk.word_tokenize(str(sentence))):
        if (pos == 'NN' or pos == 'NNP' or pos == 'NNS' or pos == 'NNPS'):
            nouns.append(word)
print(nouns)
输出:
    ['deposition', 'rate', 'leaf'...'compounds', 'inra', 'sciences']
```

以上代码中'NN'表示名词单数形式,'NNS'表示名词复数形式,'NNP'表示专有名词,'NNPS'表示专有名词复数形式。如果需要提取其他词性的词,改变词性标签即可提取。

2) NLTK 提取名词短语

读取文本后,首先用 split 函数分词,然后使用 NLTK pos_tag 标记,之后用正则表达式解析器解析标记后的文本,通过遍历列表提取名词短语。示例代码如下:

```
#导入 Python 标记库
import nltk
from nltk.tag import pos_tag

#导入 NLTK 停用词
from nltk.corpus import stopwords
stopwords = stopwords.words('english')

#定义提取名词短语函数
def extract_np(psent):                # psent==parsed sentences
    for subtree in psent.subtrees():
        if subtree.label() == 'NP':
            yield ' '.join(word for word,tag in subtree.leaves())

#定义名词短语文法
grammar = r"""
NP: {<DT|PP\$>?<JJ>*<NN>}        # chunk determiner/possessive,adjectives and noun
```

```
            {<NNP>+}           # chunk sequences of proper nouns
            {<NN>+}            # chunk consecutive nouns
    """

#命名分块解析器
cp = nltk.RegexpParser(grammar)  # chunker parser

#读取文本
sentence = open(r'CHNC-clean-para500.txt',encoding = "utf-8").read()

#标记、解析标记后的句子并提取名词短语
tagged_sent = pos_tag(sentence.split())
parsed_sent = cp.parse(tagged_sent)
for npstr in extract_np(parsed_sent):
    if npstr not in stopwords:
        print (npstr.encode('gbk','ignore').decode('gbk'))
```
输出：
herbicide deposition
rate
…
volatile compounds.
inra,
sciences
```
            print(len(npstr),end=";")
```
输出：
20;4;8;…19;5;9;

2. TextBlob 提取名词和名词短语

TextBlob 是一款自然语言处理工具，用于处理文本数据。它提供了一个简单的 API，可用于常见的自然语言处理(NLP)任务，如词性标注、名词短语提取、情感分析、分类、翻译等。TextBlob 可以轻松提取所有名词和名词短语，示例代码如下：

```
from textblob import TextBlob
File = open(r'CHNC-clean-para500.txt',encoding = "utf-8")    #open file
lines = File.read()
# print(lines)
blob = TextBlob(lines)
print(blob.noun_phrases)
```
输出：

['herbicide deposition rate', 'leaf surface',...'metabolic processes', 'volatile compounds', 'edp sciences']

3. spaCy 提取单词或短语

spaCy 是一款强大的自然语言处理工具，具有分词、词干化、情感分析、实体识别等功能，它的分词准确度高于 NLTK，速度也大大提升，可快速生成词干化并且去重的词列表。可以使用 spaCy 提取单词、短语或词对。

(1) spaCy 提取名词和形容词。示例代码如下：

```
import spacy
nlp = spacy.load('en_core_web_sm')

f_name = 'CHNC-clean-para500.txt'
f_text = open(f_name,encoding = "utf-8").read()
f_doc = nlp(f_text)
## Extract tokens for the given doc
# print ([token.text for token in f_doc])

sentences = list(f_doc.sents)
print(len(sentences))


nouns = []
adjectives = []
for token in f_doc:
    if token.pos_ == 'NOUN':
        nouns.append(token)
    if token.pos_ == 'ADJ':
        adjectives.append(token)
print(nouns[:50])
```

输出：[deposition, rate, leaf, surface, features, weeds, impact, efficacy, paper, experiment, differences, droplet, evaporation, characteristics, leaf, surface, hairy, structures, weed, lotifolium, lolium, study, leaf, structures, sprays, silicone, surfactant, concentrations, test, droplets, diameters, leaves, evaporation, characteristics, camera, evaporation, duration, droplet, coverage, images, videos, image, processing, toolbox, images, droplet, leaf, background, images&rsquo, pixel]

```
print(adjectives[:50])
```

输出：[significant, overall, orthogonal, rough, ridged, lepidium, organic, different, single, digital, maximum, binary, organic, other, small, proper, chinese, leguminous, southern, intensive, double, subsequent, early, late, moderate, early, late, total, annual, n, early, late, annual, comparable, annual, moderate, fallow, increased, main, enhanced, moderate, higher, higher, agronomic, n, n, fallow, high, sustainable, key]

如果需要提取其他词性的词，可使用表 7-1 所示的 spaCy 词性标记与之匹配。

表 7-1 spaCy 词性标记及其含义

标 记	含 义
ADJ	形容词
ADV	副词
AUX	辅助动词
CONJ	连接词
DET	限定词
INTJ	感叹词
NOUN	名词
NUM	数字
PART	小品词
PRON	代词
PROPN	专有名词
PUNCT	标点符号
SCONJ	从属连词
SYM	符号
VERB	动词

(2) spaCy 提取名词短语。示例代码如下：

```
import spacy
nlp = spacy.load('en_core_web_sm')

f_name = 'CHNC-clean-para500.txt'
f_text = open(f_name,encoding = "utf-8").read()
f_doc = nlp(f_text)

for np in f_doc.noun_chunks:
    print(np.text, np.root.dep_, np.root.head.text)
```
输出：herbicide deposition rate nsubjpass affected

the leaf surface features pobj by

weeds pobj of

a significant impact dobj have

…

volatile compounds pobj of

inra ROOT inra

edp sciences appos inra

(3) spaCy 提取名词、形容词对。示例代码如下：

```python
import spacy
nlp = spacy.load('en_core_web_sm')

f_text = open('CHNC-clean-para500.txt',encoding = "utf-8").read()
f_doc = nlp(f_text)

n_adj_pairs = []
for i,token in enumerate(f_doc):
    if token.pos_ not in ('NOUN','PROPN'):
        continue
    for j in range(i+1,len(f_doc)):
        if f_doc[j].pos_ == 'ADJ':
            n_adj_pairs.append((token,f_doc[j]))
            break
print(n_adj_pairs)
```
输出：[(herbicide, significant), (deposition, significant), (rate, significant) ... (regulators, metabolic), (processes, volatile), (formation, volatile)]
```python
print(len(n_adj_pairs))
```
输出：51790

7.3 提取句型

spaCy 是功能强大的自然语言处理库，可用于从文本中提取特定句型。提取句型的基步骤是进行依存句法分析，标注出句子中的词语依存关系，然后根据依存关系的标签判断句子的语法结构。利用结构信息，建立规则判断句子是否匹配给定句型模板。具体提取方法如下：

(1) 导入 spaCy 并加载模型，对文本分词，标注依存关系。
(2) 定义判断规则函数，根据依存标签检查句型模板的关键元素是否存在。
(3) 遍历文本的句子，使用判断函数过滤出匹配的句子。
(4) 定义不同的规则函数，提取不同句型。

1. 提取动词 + 副词句型

用 spaCy 提取动词 + 副词句型的句型结构。示例代码如下：

```python
import spacy
from spacy.matcher import Matcher
from spacy.util import filter_spans

nlp = spacy.load('en_core_web_sm')

f_name = 'CHNC-clean-para500.txt'
```

```python
f_text = open(f_name,encoding = "utf-8").read()
print(type(f_text))


pattern = [{'POS': 'VERB', 'OP': '+'}, ### verb + adv
           {'POS': 'ADV', 'OP': '+'}] ## 1,or more

# instantiate a Matcher instance
matcher = Matcher(nlp.vocab)
matcher.add("Verb phrase", None, pattern)

doc = nlp(f_text)

# call the matcher to find matches
matches = matcher(doc)
spans = [doc[start:end] for _, start, end in matches]

print (filter_spans(spans))
```
输出：[spread more widely, evaporated faster, varied much faster,… achieved only, suggested accordingly, can therefore]

改变以上代码中的 pattern 即可提取不同的句型结构，例如提取副词 + 动词，使用代码 pattern = [{'POS': 'ADV', 'OP': '+'}, {'POS': 'VERB', 'OP': '+'}]，提取所有副词修饰动词的情况，不论副词在动词前面还是后面，使用代码 pattern = [{'POS': 'VERB', 'OP': '?'}, {'POS': 'ADV', 'OP': '*'}, {'POS': 'VERB', 'OP': '+'}]即可提取。

2. 提取 SVO 句型

用 spacy 提取 SVO 句型结构。示例代码如下：

```python
# 导入 spacy 模块
import spacy

# 从符号模块导入依存关系标签
from spacy.symbols import nsubj, VERB, dobj

# 加载英文语言模型
nlp = spacy.load("en_core_web_sm")

# 打开文本文件并读取内容
with open('CHNC-clean-para500.txt', 'r',encoding = 'utf-8') as file:
    text = file.read()

# 处理文本，获取 Doc 对象
```

```
doc = nlp(text)

# 定义列表，存储提取的 SVO 句型
svo_patterns = []

# 遍历每个词，找出符合 SVO 的短语
for possible_subject in doc:
    if possible_subject.dep == nsubj and possible_subject.head.pos == VERB:
        for possible_object in possible_subject.head.children:
            if possible_object.dep == dobj:
                # 提取主语、动词、宾语文本，添加到列表
                svo_patterns.append((possible_subject.text, possible_subject.head.text, possible_object.text))

# 打印提取的 SVO 句型
for pattern in svo_patterns:
    print(f"主语: {pattern[0]}, 动词: {pattern[1]}, 宾语: {pattern[2]}")

# 将 SVO 句型保存到文件
with open('extracted-SVO.txt', 'w', encoding='utf-8') as f:
    for pattern in svo_patterns:
        f.write(f"主语: {pattern[0]}, 动词: {pattern[1]}, 宾语: {pattern[2]}\n")
```

运行代码后，代码将提取的 SVO 句型保存到名为 "extracted-SVO.txt" 的文件中并打印出 SVO 句型结果。部分输出内容如下：

主语: we，动词: conducted，宾语: experiment

7.4 提取特定特征文本

1. 提取自定义结构

利用正则表达式与分块结合提取自定义结构主要是通过构建语法模式，用正则表达式解析器识别块结构，最后用遍历方法通过 label 提取自定义结构。其中语法模式也可包括多条规则，提取结构时用 "or" 连接多个可选条件，最后一次性输出自定义结构。提取自定义结构代码示例如下：

```
#导入 nltk 和分词器
import nltk
from nltk import sent_tokenize
#读取文本
raw_text = open('CHNC-clean-para500N.txt',encoding = "utf-8").read()
sentences = nltk.sent_tokenize(raw_text)
```

```
sentences = [s.split() for s in sentences]
#词性标注
sentences = [nltk.pos_tag(s) for s in sentences]
#构建语法模式
pattern = "'PSV:{<MD><VB><VBN>+}'"
#识别块结构
Chunker = nltk.RegexpParser(pattern)
for s in sentences:
    chunks = Chunker.parse(s)
    # print(chunks)
#提取自定义块结构并输出
    for subtree in chunks.subtrees():
        if subtree.label() == 'PSV':
            print(subtree)
```
输出：(PSV can/MD be/VB affected/VBN)

(PSV could/MD be/VB captured/VBN)

(PSV can/MD be/VB combined/VBN)

…

(PSV could/MD be/VB achieved/VBN)

(PSV will/MD be/VB achieved/VBN)

(PSV can/MD be/VB used/VBN)

如果需要匹配多个句型结构，上面代码中的 pattern 可以改为 pattern = r"""。

```
    NP: {<DT|JJ|NN.*>+}          # 限定词或形容词+名词
    PP: {<IN><NP>}               # 介词+名词
    VP: {<VB.*><NP|PP|CLAUSE>+$} # 动词+名词或介词或从句
    CLAUSE: {<NP><VP>}           # 名词+动词
    """
```

遍历分块时代码如下：

```
for subtree in chunks.subtrees():
    if subtree.label() == 'VP'     or subtree.label() == 'PP'    or subtree.label() == 'NP'    or subtree.label() == 'CLAUSE':
        print(subtree)
```

输出：NP herbicide/JJ deposition/NN rate/NN)

(PP by/IN (NP the/DT leaf/NN surface/NN features/NNS))

(NP the/DT leaf/NN surface/NN features/NNS)

…

(NP the/DT formation/NN)

(PP of/IN (NP volatile/JJ compounds./NN))

(NP volatile/JJ compounds./NN)

2. 提取含特定词的完整句子

(1) 使用 NLTK 分句、分词后提取句子。示例代码如下：

```
from nltk.tokenize import sent_tokenize
from nltk.tokenize import word_tokenize
import codecs

def sentence_finder(text,word):
    sentences=sent_tokenize(text)
    return [sent for sent in sentences if word in word_tokenize(sent)]

f = open(r'filepath\filename.txt', encoding = "gbk").read()

sent = sentence_finder(f,'but')
print(sent)
print(len(sent))
```

(2) 使用 split()方法结合遍历提取含有某个单词的句子。示例代码如下：

```
sentences = open(r'filepath\filename.txt', encoding = "gbk").readlines()

sentences=str(sentences).split('.')
#遍历所有分句，查找含有 which 的句子
sentences=[sent for sent in sentences if 'which' in sent]
print (len(sentences))
输出：168
for sent in sentences:
    print(sent+'.')
```

例子中提取的句子基本正确，但个别句子仍含有多余部分，主要原因是使用了"."分句，不能校正原文本中问题句子。如果原文本中含有非句子结尾处的"."，分句时也会被认为是句子结尾。要解决这一问题，可以通过加载停用词等方式处理。

(3) 使用正则表达式提取含自定义单词或字符的句子。示例代码如下：

```
#读取文本
file = open(r'filepath\filename.txt', encoding = "gbk").read()
#分句
sentences=file.split('.')
#列表转为字符串并换行
str_sentences = '\n'.join(sentences)
#导入正则表达式库
import re
#匹配以 this 开头并以 area 为结尾的句子
pattern = 'this[ ](.*?)area'
```

```
#正则表达式编译、查找
sents_found=re.compile(pattern).findall(str_sentences)
#遍历查找结果并输出
for sent in sents_found:
    print(sent)
```

(4) 使用 startswith()和 endswith()函数提取以特定词或字符开始、结尾的句子。示例代码如下：

```
file = open(r'filepath\filename.txt', encoding = "gbk").read()

sentences=file.strip( ).split('.')
#遍历所有句子提取以 this 开头并以 characteristics 结尾的句子
for line in sentences:
    if line.startswith(" this") and line.endswith("characteristics"):
        print(line)
```

注意：该函数对行首空格敏感，空格会影响匹配结果。

(5) 使用 contains()方法提取包含特定字符串的信息。示例代码如下：

```
import pandas as pd
import csv
path = r"filepath\filename.csv"    # csv 文件路径
df = pd.DataFrame(pd.read_csv(path,encoding = "utf-8"))   # 读取 csv 文件
df.columns = ['fetchedDate','NewDate','pub_date','tid', 'title','parse_content'] #命名列名称
df['title'] = df['title'].astype(str)# "title" # 列作为字符串类型
dataR_1 = df[df['title'].str.contains('数据|资源')] #提取"title"列中含有"数据"或"资源"的信息
print(len(dataR_1)) #查看信息数
输出：1023
otherINFO = df[~df['title'].str.contains('数据|资源')] #提取"title"列中不含"数据"或"资源"的信息
print(len(otherINFO)) #查看信息数
输出：2090
otherINFO.to_csv('数据资源以外其他信息 MAR-4.csv', encoding='utf_8_sig')# 保存提取
```

到的不含"数据"或"资源"的信息为 CSV 文件，文件保存在本程序所在的文件夹内。

已保存部分文件截图如图 7-1 所示。

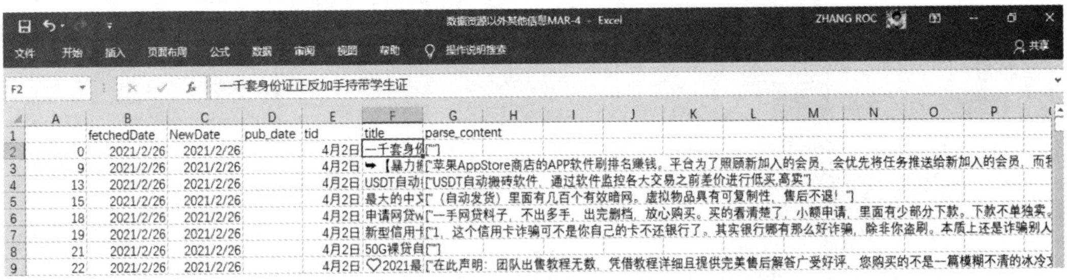

图 7-1 已保存部分文件截图

7.5 提取关键词

关键词提取是从文本中检索关键词或关键短语的过程，可以利用关键词或关键短语概括文档的主题。提取关键词可分为有监督方法提取和无监督方法提取两种。

(1) 有监督方法提取分为需要明确标注和不需要明确标注两种情况。需要明确标注的方法包括基于二分类的方法，如 SVM、Logistic 回归等分类算法；基于序列标注的方法，如双向长短时记忆网络(BI-LSTM)等。不需要明确标注的方法是让模型自动学习关注关键词的基于注意力机制的方法。

(2) 无监督方法提取关键词无需人工标注，常见的有基于统计的方法，如根据词频和逆文档频率(TF-IDF)确定关键词；基于图模型的 TextRank 方法，构建图模型分析词语的重要程度自动计算关键词；基于主题模型的方法，通过分析文档的主题分布来提取关键词，如潜在狄利克雷分布(Latent Dirichlet Allocation，LDA)等。此外，还有基于词向量的方法、大语言模型的方法。

本节主要介绍 TF-IDF、TextRank、LDA 模型方法、词向量与 LDA 相结合的方法和 Openai GPT 模型提取关键词。在实际应用中，需要根据具体的文本特征和任务需求来权衡使用各种方法的利弊。

1. TF-IDF + TextRank

TF-IDF 与 TextRank 是经典的关键词提取算法，以下载报纸文本为例，用 jieba 分词后提取关键词。代码如下：

```
# 导入 jieba 分词、关键词提取模块、Pandas 库和 newspaper 库
import jieba.analyse
import pandas as pd
import newspaper

# 定义文章的 URL 地址
url = 'https://finance.sina.com.cn/roll/2021-01-23/doc-ikftpnny0756403.shtml'

# 创建一个新闻文章对象，设置语言为中文
article = newspaper.Article(url, language='zh')

# 下载并解析文章
article.download()
article.parse()

# 检查文章是否下载和解析成功
if article.download_state == 2 and article.is_parsed:

    article.nlp()
```

```python
        string_data = "".join(article.keywords)

        # 定义提取关键词的函数
        def get_key_words(string_data, method=''):
            # 如果方法是'textrank',则使用 TextRank 算法,否则,默认使用 TF-IDF 算法
            if method == 'textrank':
                tags_pairs = jieba.analyse.textrank(string_data, topK=5, withWeight=True)
            else:
                tags_pairs = jieba.analyse.extract_tags(string_data, topK=5, withWeight=True)

            # 创建一个包含单词及其权重的元组列表
            tags_list = [(i[0], i[1]) for i in tags_pairs]

            # 将列表转换成 DataFrame
            tags_pd = pd.DataFrame(tags_list, columns=['word', 'weight'])

            return tags_pd

        # 使用 TF-IDF 提取关键词
        TF_keywords = get_key_words(string_data)

        # 使用 Textrank 提取关键词
        Textrank_keywords = get_key_words(string_data, method='textrank')

        # 打印提取的关键词
        print("TF-IDF 关键词:\n", TF_keywords)
        print("Textrank 关键词:\n", Textrank_keywords)
else:
        print("文章下载或解析失败。")
```

运行以上代码,输出权重最大的 5 个关键词和它们的权重值,部分运行结果如下:

```
TF-IDF 关键词:
      word    weight
0     负债    0.803071
1   商业银行   0.297447
2   业务管理   0.163995
Textrank 关键词:
      word    weight
0     负债    1.000000
1     质量    0.295957
2     管理    0.257976
```

TF-IDF 与 TextRank 的结合是一种强大的关键词提取方法，综合了两种算法的优势。这种组合方法的主要优点是能够同时考虑词频统计信息和词语之间的语义关系。TF-IDF 提供了词语在文档中的重要性和区分度的量化指标，而 TextRank 则通过图模型捕捉了词语之间的上下文联系。这种结合可以更全面地评估词语的重要性，既考虑了词语的频率，又不忽视语义上的重要性，从而能提取出更加准确和有代表性的关键词。

这种组合方法也存在一些缺点。首先，它的计算复杂度相对较高，尤其是在处理大规模文本数据时会面临效率低的问题。其次，这种方法会受到参数设置的影响，如 TextRank 中的窗口大小、迭代次数等，需要经过多次调整才能获得最佳效果。此外，对于一些特殊类型的文本，如非常短的文本或者专业术语密集的文档，这种组合方法可能不如单独使用 TF-IDF 或其他专门的方法效果好。

另外，这种方法可能会在某些情况下过度强调某些词语。例如，如果一个词在 TF-IDF 中得分很高，同时在文本中与其他重要词语有很多连接，那可能会在最终结果中被过度强调，而忽视了一些同样重要但不那么显著的词语。

总的来说，TF-IDF 和 TextRank 的组合方法在大多数情况下能够提供很好的关键词提取效果，特别是对于需要同时考虑统计信息和语义关系的场景。

2. BoW + LDA 主题模型

BoW + LDA 主题模型提取关键词的方法先使用词袋模型(Bag of Words，BoW)处理文本数据，然后利用 LDA 主题模型提取关键词。词袋模型是一种简单的文本表示方法，它将文本转换为词的集合，不考虑语法和词序。这种方法提取主题关键词的主要过程包括分词、构建词典和向量化、使用处理后的词袋向量训练 LDA 模型、识别主题和提取关键词。LDA 将每个文档表示为一系列主题的混合，同时每个主题表示为一系列关键词的混合。这个过程涉及确定每个文档对应的主题分布，以及每个主题下词的分布。关键词提取根据每个主题下的词分布，确定对主题贡献最大的词为关键词。

以提取"CHNC-clean-para500.txt"文本主题为例，使用 BOW 处理文本，训练 LDA 主题模型提取文本主题关键词，保存关键词在名为"BowLdaKeywds.txt"文件中并打印出结果。代码如下：

```
# 导入 LDA 主题模型库 Gensim、NLTK 停用词和其他必要的库
import gensim
from gensim import corpora
from nltk.corpus import stopwords
from collections import defaultdict
import nltk

# 下载 NLTK 停用词
nltk.download('stopwords')

# 打开并读取文件
with open('CHNC-clean-para500.txt', 'r', encoding='utf-8') as file:
```

```
    text = file.read()

# 分词并移除停用词
stop_words = set(stopwords.words('english'))
texts = [[word for word in document.lower().split() if word not in stop_words]
        for document in text.split('\n')]

# 过滤出现次数少于 2 次的单词
frequency = defaultdict(int)
for text in texts:
    for token in text:
        frequency[token] += 1

texts = [[token for token in text if frequency[token] > 1] for text in texts]

# 创建字典(词袋)
dictionary = corpora.Dictionary(texts)

# 预设 5 个话题，训练 LDA 模型
corpus = [dictionary.doc2bow(text) for text in texts]
lda_model = gensim.models.LdaModel(corpus, num_topics=5, id2word=dictionary, passes=15)

# 提取并保存关键词
topics = lda_model.print_topics(num_topics=5, num_words=10)
keywords = [', '.join(word.split('*')[1].replace('"', '') for word in topic[1].split('+')) for topic in topics]

with open('BowLdaKeywds.txt', 'w', encoding='utf-8') as f:
    for topic_keywords in keywords:
        f.write(topic_keywords + '\n')

# 打印主题
for topic in topics:
    print(topic)
```

运行代码，输出结果中第一个主题关键词如下：

(0, '0.015*"soil" + 0.015*"n" + 0.011*"yield" + 0.010*"rice" + 0.007*"p" + 0.007*"increased" + 0.007*"application" + 0.006*"fertilizer" + 0.006*"maize" + 0.006*"compared"')

BoW 结合 LDA 是一种较为先进的关键词提取方法。这种组合的主要优点是能够捕捉文档的主题结构，同时考虑词语在不同主题下的分布。LDA 可以发现潜在的主题，而 BoW 提供了文档的词频信息，两者结合可以更好地识别出与文档主题相关的关键词。这种方法

特别适用于长文本或者具有多个主题的文档集合，能够提取出更加与语义相关和主题相关的关键词。

这种方法也存在一些缺点。首先，LDA 需要预先设定主题数量，这个参数的选择可能会影响结果质量。如果主题数量设置不当，可能会导致关键词提取不准确。其次，LDA 是一个概率模型，其结果具有一定的随机性，多次运行可能会得到略有不同的结果。此外，BoW + LDA 方法计算复杂度较高，特别是在处理大规模文档集时，可能会面临效率低的问题。还有，这种方法忽略了词序和上下文信息，可能会遗漏一些重要但频率不高的词语。对于短文本或者专业性很强的文档，BoW + LDA 可能不如其他更简单的方法(如 TF-IDF)效果好。

总的来说，BoW + LDA 在提取主题相关的关键词方面具有独特优势，特别适用于需要深入理解文档主题结构的场景。

3. TF-IDF + LDA 主题模型

以提取"CHNC-clean-para500.txt"文本主题为例，使用 TF-IDF 算法处理文本，训练 LDA 主题模型提取文本主题关键词，保存关键词在名为"BowLdaKeywds.txt"文件中并打印出结果。代码如下：

```python
# 导入所需的库
import gensim
from sklearn.feature_extraction.text import TfidfVectorizer
from nltk.corpus import stopwords
import nltk
import os

# 下载 NLTK 停用词
nltk.download('stopwords')

# 打开并读取文件
with open('CHNC-clean-para500.txt', 'r', encoding='utf-8') as file:
    text = file.readlines()

# 对文本进行分词并移除停用词
stop_words = set(stopwords.words('english'))
texts = [' '.join([word for word in document.lower().split() if word not in stop_words])
        for document in text]

# 过滤掉低频词
vectorizer = TfidfVectorizer(min_df=2)
tfidf_matrix = vectorizer.fit_transform(texts)

# 将稀疏矩阵转换为 gensim 语料库
```

```
corpus = gensim.matutils.Sparse2Corpus(tfidf_matrix, documents_columns=False)

# 训练 LDA 模型
dictionary = gensim.corpora.Dictionary.from_corpus(corpus, id2word=dict((id, word) for word, id in vectorizer.vocabulary_.items()))
lda_model = gensim.models.LdaModel(corpus=corpus, num_topics=5, id2word=dictionary, passes=15)

# 提取并保存关键词
topics = lda_model.print_topics(num_topics=5, num_words=10)
keywords = [', '.join(word.split('*')[1].replace('"', '') for word in topic[1].split('+')) for topic in topics]

with open('TfidfLdaKeywds.txt', 'w', encoding='utf-8') as f:
    for topic_keywords in keywords:
        f.write(topic_keywords + '\n')

# 打印主题
for topic in topics:
    print(topic)
```

运行代码，输出结果中第一个主题关键词如下：

(0, '0.003*"droplet" + 0.001*"uav" + 0.001*"nozzles" + 0.001*"spray" + 0.001*"deposition" + 0.001*"droplets" + 0.001*"drift" + 0.001*"nozzle" + 0.001*"chestnut" + 0.001*"processing"')

TF-IDF 与 LDA 结合的关键词提取方法融合了统计信息和主题模型的优势。其主要优点是能够同时考虑词语的文档频率和主题分布。TF-IDF 提供了词语在文档集合中的重要性和独特性的量化指标，LDA 则能够发现文档的潜在主题结构。这种组合可以更全面地评估词语的重要性，既考虑了词语在整个文档集中的分布特征，又能识别出与特定主题相关的关键词。特别是在处理大规模、多主题的文档集合时，这种方法能够提取出更加与语义相关和主题相关的关键词。

然而，这种组合方法的计算复杂度较高，特别是 LDA 部分，在处理大规模数据时可能会面临效率低的问题。LDA 需要预先设定主题数量，这个参数的选择可能会显著影响结果的质量，需要通过多次实验或者使用困惑度、一致性分析来确定最佳的主题数。此外，LDA 结果的随机性可能会影响关键词提取的稳定性。

在某些情况下，这种方法可能会忽视一些重要但频率不高的词语。例如，如果一个重要的专业术语在整个文档集中出现频率不高，它可能在 TF-IDF 评分中得分较低，同时在 LDA 的主题分布中也不够突出，从而在最终的关键词列表中被忽视。

总的来说，TF-IDF 和 LDA 的组合方法在处理大规模、多主题文档集时表现出色，能够提取出既统计显著又与主题相关的关键词。

4. Word2Vec + LDA 方法提取关键词

以提取"CHNC-clean-para500.txt"文本主题为例，使用 Word2Vec 和 LDA 主题模型提取文本主题关键词，保存关键词在名为"Wd2VecLdaKeywds.txt"文件中并打印出结果：

(1) 使用 Word2Vec 进行文本分析和聚类。代码如下：

```python
#导入所需的库
import gensim
from nltk.corpus import stopwords
from nltk.tokenize import word_tokenize
import nltk
import numpy as np
from sklearn.cluster import KMeans

# 下载所需的 NLTK 停用词
nltk.download('punkt')
nltk.download('stopwords')

# 读取文本数据
with open('CHNC-clean-para500.txt', 'r', encoding='utf-8') as file:
    text = file.read().lower()

# 分词和移除停用词
stop_words = set(stopwords.words('english'))
words = [word for word in word_tokenize(text) if word.isalpha() and word not in stop_words]

# 训练 Word2Vec 模型
model = gensim.models.Word2Vec([words], vector_size=100, window=5, min_count=2, workers=4)

# 检查模型是否训练了词向量
if not model.wv:
    raise ValueError("The Word2Vec model didn't train any word vectors.")

# 定义聚类的数量
num_clusters = 5

# 使用 KMeans 进行聚类
word_vectors = model.wv.vectors

# 测试 KMeans 聚类部分
try:
    kmeans = KMeans(n_clusters=num_clusters)
    idx = kmeans.fit_predict(word_vectors)
```

```
except AttributeError as e:
    print(f"AttributeError occurred: {e}")
    raise
except Exception as e:
    print(f"An unexpected error occurred: {e}")
    raise


# 获取每个聚类的关键词
word_centroid_map = dict(zip(model.wv.index_to_key, idx))
clustered_words = [[] for _ in range(num_clusters)]
for word, cluster in word_centroid_map.items():
    clustered_words[cluster].append(word)


# 选择每个聚类中的前 N 个词作为特征
N = 10
selected_words = [cluster[:N] for cluster in clustered_words]


# 打印每个聚类的前 N 个词
for i, words_in_cluster in enumerate(selected_words):
    print(f"Cluster {i + 1}: {words_in_cluster}")
```

(2) 使用 LDA 进行主题模型分析。代码如下：

```
from gensim import corpora


# 使用选定的词汇构建 LDA 模型
dictionary = corpora.Dictionary(selected_words)
corpus = [dictionary.doc2bow(text) for text in selected_words]
lda_model = gensim.models.LdaModel(corpus, num_topics=5, id2word=dictionary, passes=15)


# 提取并打印 LDA 主题
topics = lda_model.print_topics(num_words=10)
for topic in topics:
    print(topic)
```

(3) 融合 Word2Vec 和 LDA 的结果，打印并保存到文件。代码如下：

```
with open('Wd2VecLdaKeywds.txt', 'w', encoding='utf-8') as file:
    for topic_num, topic in enumerate(lda_model.print_topics(num_topics=5, num_words=20)):
        # 提取每个主题的前 20 个关键词
        words = [word.split('*')[1].replace('"', '').strip() for word in topic[1].split('+')][:20]
```

```
    # 打印主题
    print(f"Topic {topic_num + 1}:")
    file.write(f"Topic {topic_num + 1}:\n")

    # 选择每个主题的前 10 个关键词作为最终关键词
    final_keywords = words[:10]
    for word in final_keywords:
        print(f"{word}")
        file.write(f"{word}\n")

    print("\n")
    file.write("\n")
```

运行以上代码,程序会分别打印出(1)(2)(3)的结果,同时,也会把 5 个主题下的关键词保存在名为"Wd2VecLdaKeywds.txt"文件中。

代码使用 Word2Vec 对文本进行初步分析,通过 KMeans 聚类找到相关的词汇。接着,这些词汇被用作特征构建 LDA 模型。然后,在结果层面进行融合,通过比较 Word2Vec 的语义相似性和 LDA 的主题分布,提取出更加丰富的关键词或概念。这样,就能够有效地利用 Word2Vec 和 LDA 的优势,提取出文本的关键信息。

5. GloVe + LDA 主题模型

以提取 "CHNC-clean-para500.txt" 文本主题为例,使用 GloVe 向量和 LDA 主题模型提取文本主题关键词,保存关键词在名为 "GloveLdaKeywds.txt" 文件中并打印出结果。代码如下:

```
#导入所需的库
import spacy
import numpy as np
from sklearn.cluster import KMeans
from gensim import corpora
from gensim.models.ldamodel import LdaModel


def load_glove_vectors(glove_file):
    """ 从文件中加载 GloVe 向量 """
    word_vectors = {}
    try:
        with open(glove_file, 'r', encoding='utf-8') as f:
            for line in f:
                parts = line.split()
                word_vectors[parts[0]] = np.array(parts[1:], dtype=np.float32)
    except FileNotFoundError:
```

```python
            print(f"File not found: {glove_file}")
    return word_vectors


def preprocess_text(file_path, nlp):
    """ 使用 SpaCy 进行文本预处理 """
    try:
        with open(file_path, 'r', encoding='utf-8') as file:
            text = file.read().lower()
    except FileNotFoundError:
        print(f"File not found: {file_path}")
        return []

    doc = nlp(text)
    return [token.text for token in doc if not token.is_stop and not token.is_punct]


def print_and_save_results(filename, lda, dictionary):
    """ 打印并保存 LDA 结果 """
    try:
        with open(filename, 'w', encoding='utf-8') as file:
            for topic_num in range(lda.num_topics):
                top_terms = lda.get_topic_terms(topic_num, topn=15)
                topic_info = f"Topic {topic_num + 1}:\n"
                print(topic_info)
                file.write(topic_info)

                for term_id, prob in top_terms:
                    word = dictionary[term_id]
                    keyword_with_weight = f"{word}: {prob:.4f}\n"
                    print(keyword_with_weight, end='')
                    file.write(keyword_with_weight)

                file.write("\n")
                print("\n")
    except IOError:
        print(f"Error writing to file: {filename}")


# 初始化 SpaCy 并加载 GloVe 向量
nlp = spacy.load("en_core_web_sm")
```

```python
glove_vectors = load_glove_vectors('H:/BaiduNetdiskDownload/glove.6B/glove.6B.100d.txt')

# 预处理文本
tokens = preprocess_text('CHNC-clean-para500.txt', nlp)
filtered_tokens = [word for word in tokens if word in glove_vectors]

# 为过滤后的 token 创建向量
vectors = [glove_vectors[word] for word in filtered_tokens]

# K-Means 聚类
kmeans = KMeans(n_clusters=10)
idx = kmeans.fit_predict(vectors)

# 构建包含 10 个主题的 LDA 模型
clustered_words = [filtered_tokens[i] for i in idx]
dictionary = corpora.Dictionary([clustered_words])
corpus = [dictionary.doc2bow(clustered_words)]
lda = LdaModel(corpus, num_topics=10)

# 打印并保存 10 个主题及其每个主题的 15 个关键词
print_and_save_results('GloveLdaKeywds.txt', lda, dictionary)
```

运行以上代码，保存并打印提取到的 10 个主题的前 15 个权重较高关键词及其权重值。

以上代码使用 spaCy 预处理文本，结合了 GloVe 向量和 LDA 主题建模，通过聚类分析和主题提取，提取了文本关键词用于探索和理解文本数据中的潜在主题。这种方法可以应用于文本挖掘、信息检索、文档分类等多种场景。

6. LDA 主题模型关键词可视化及验证

(1) LDA 主题模型提取关键词可视化。以提取文件 "imdb_review-s.csv" 中 "review" 列主题关键词为例，加载 NLTK 停用词和自定义停用词预处理文本后用 pyLDAvis 可视化主题关键词。代码如下：

```python
# 导入所需的库
import pyLDAvis
import pandas as pd
import nltk
from nltk.corpus import stopwords
from nltk.tokenize import RegexpTokenizer
from nltk.stem.wordnet import WordNetLemmatizer
from gensim.corpora import Dictionary
```

```python
from gensim.models import LdaModel
import pyLDAvis.gensim_models as gensimvis
import matplotlib.pyplot as plt

# 确保 NLTK 资源可用
nltk.download('stopwords')
nltk.download('wordnet')

# 从 NLTK 获取英文停用词集
stop_words_en = set(stopwords.words('english'))

# 配置 matplotlib 以正确显示中文字符
plt.rcParams['font.sans-serif'] = ['SimHei']
plt.rcParams['axes.unicode_minus'] = False

def get_custom_stopwords(file_path):
    """从文件中加载自定义停用词。"""
    custom_stop_words = set()
    try:
        with open(file_path, 'r') as file:
            custom_stop_words = set(word.strip() for word in file)
    except FileNotFoundError:
        print(f"在路径：{file_path}未找到自定义停用词文件")
    return custom_stop_words

def preprocess_text(text, tokenizer, lemmatizer, stop_words):
    """清洗和预处理文本。"""
    if not isinstance(text, str):
        return []
    text = text.lower()
    tokens = tokenizer.tokenize(text)
    tokens = [token for token in tokens if not token.isnumeric() and len(token) > 1]
    tokens = [token for token in tokens if token not in stop_words]
    tokens = [lemmatizer.lemmatize(token) for token in tokens]
    return tokens

def get_corpus_dictionary(df, stop_words):
    """从数据框生成语料库和字典。"""
```

```python
        tokenizer = RegexpTokenizer(r'\w+')
        lemmatizer = WordNetLemmatizer()

        texts = [preprocess_text(document, tokenizer, lemmatizer, stop_words) for document in df['review'] if
isinstance(document, str)]

        dictionary = Dictionary(texts)
        corpus = [dictionary.doc2bow(text) for text in texts]

        print('唯一标记的数量：%d' % len(dictionary))
        print('文档的数量：%d' % len(corpus))

        return corpus, dictionary

def test_lda():
    """测试并将 LDA 可视化保存为 HTML。"""
    # 定义自定义停用词文件的绝对路径
    custom_stopwords_path = "./phstopwds.txt"
    custom_stop_words = get_custom_stopwords(custom_stopwords_path)
    combined_stop_words = stop_words_en | custom_stop_words

    # 加载数据集并确保正确的编码
    df = pd.read_csv("imdb_review-s.csv", encoding='Latin-1')
    corpus, dictionary = get_corpus_dictionary(df, combined_stop_words)

    # 创建并训练 LDA 模型
    lda = LdaModel(corpus=corpus, num_topics=15)
    data = gensimvis.prepare(lda, corpus, dictionary)

    # 将 LDA 可视化保存为 HTML 文件
    pyLDAvis.show(data, open_browser=True, local=False)
    pyLDAvis.save_html(data, 'LdavisImdb.html')

if __name__ == "__main__":
    test_lda()
```

运行以上代码，pyLDAvis 生成交互式的 LDA 主题关键词可视化如图 7-2 所示。图中每个主题被表示为一个气泡，气泡的大小表示主题的重要性，气泡之间的距离表示主题之间的相似性。可以向左、向右拖动右上方的 λ 值调整主题关键词，选择适当的关键词。

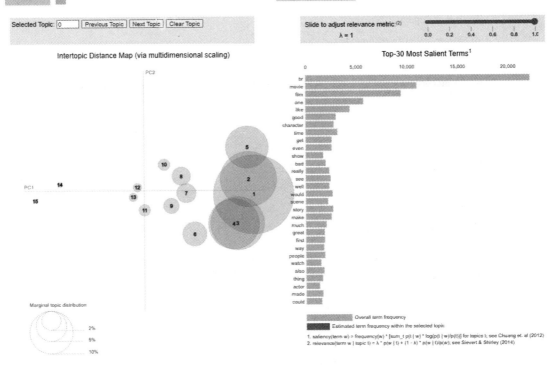

图 7-2 LDA 主题关键词可视化

(2) LDA 主题模型困惑度和一致性验证。通过计算模型的困惑度(Perplexity)或一致性(Coherence)来验证模型的效果，评估主题关键词的可信度。困惑度越低，一致性越高，说明模型的效果越好。当主题数很多时，生成的模型可能会过拟合，需要结合一致性值判断模型可靠性。在实际应用中，通常会先计算不同主题数量下模型的困惑度，然后在困惑度较低的范围内，选择一致性最高的主题数量作为最终的主题数量，这种情况下生成的关键词可信度较高。

以提取文件"imdb_review-s.csv"主题并验证主题困惑度和一致性为例，使用 spaCy，Pandas、NLTK、Gensim，和 Matplotlib 等库来进行文本预处理、数据分析和可视化。首先检查 NLTK 停用词和词性标注资源，然后将文本转换为小写，分词，移除数字和 NLTK 停用词以及自定义停用词"phstopwds.txt"，并用 WordNet 进行词形还原，用 Gensim 提取文本，应用二元模型和三元模型，进而用 spaCy 从处理后的数据中提取特定词性的单词的基本形式，创建、优化词典和语料库，构建和评估 LDA 模型，保存并打印 LDA 模型的每个主题的关键词及其权重，最后计算模型的困惑度、一致性，可视化困惑度、一致性并保存困惑度、一致性值及其图示。代码如下：

```
# 导入程序所需的库
import spacy
from multiprocessing.dummy import freeze_support
import re
import pandas as pd
import nltk
from nltk.corpus import stopwords
```

```python
from nltk.tokenize import RegexpTokenizer
from nltk.stem.wordnet import WordNetLemmatizer
import gensim
import gensim.corpora as corpora
from gensim.corpora import Dictionary
from gensim.models import LdaModel, CoherenceModel
import matplotlib
matplotlib.use('TkAgg')
import matplotlib.pyplot as plt

# 检查停用词和词性标注资源
def check_nltk_resources():
    for resource in ['corpora/stopwords', 'corpora/wordnet']:
        try:
            nltk.data.find(resource)
        except LookupError:
            nltk.download(resource.split('/')[1])

check_nltk_resources()


# 设置 matplotlib 以正确显示中文字符
plt.rcParams['font.sans-serif'] = ['SimHei']
plt.rcParams['axes.unicode_minus'] = False


# 从文件加载自定义停用词
def get_custom_stopwords(file_path):
    try:
        with open(file_path, 'r') as file:
            return set(word.strip() for word in file)
    except FileNotFoundError:
        print(f"自定义停用词文件未找到: {file_path}")
        return set(stopwords.words('english'))

# 清洗并预处理文本
def preprocess_text(text, tokenizer, lemmatizer, stop_words):
    if not isinstance(text, str):
        return []
    text = text.lower()
    tokens = tokenizer.tokenize(text)
    tokens = [token for token in tokens if not token.isnumeric() and len(token) > 1]
```

```
        tokens = [token for token in tokens if token not in stop_words]
        tokens = [lemmatizer.lemmatize(token) for token in tokens]
        return tokens

    # 从数据集中获取文本、构建字典和语料库
    def get_texts_dict_corpus():
        custom_stopwords_path = "./phstopwds.txt"
        custom_stop_words = get_custom_stopwords(custom_stopwords_path)
        combined_stop_words = set(stopwords.words('english')) | custom_stop_words

        df = pd.read_csv("imdb_review-s.csv", encoding='Latin-1')
        data = df['review'].astype(str)

        tokenizer = RegexpTokenizer(r'\w+')
        lemmatizer = nltk.stem.WordNetLemmatizer()
        data_words = [preprocess_text(doc, tokenizer, lemmatizer, combined_stop_words) for doc in data]

        bigram = gensim.models.Phrases(data_words, min_count=5, threshold=100)
        trigram = gensim.models.Phrases(bigram[data_words], threshold=100)
        bigram_mod = gensim.models.phrases.Phraser(bigram)
        trigram_mod = gensim.models.phrases.Phraser(trigram)

        data_trigrams = [trigram_mod[bigram_mod[doc]] for doc in data_words]

        nlp = spacy.load('en_core_web_sm', disable=['parser', 'ner'])
        data_lemmatized = [[token.lemma_ for token in nlp("".join(sent)) if token.pos_ in ['NOUN', 'ADJ', 'VERB', 'ADV']]
                           for sent in data_trigrams]

        id2word = corpora.Dictionary(data_lemmatized)
        id2word.filter_extremes(no_below=5, no_above=0.5)

        corpus = [id2word.doc2bow(text) for text in data_lemmatized]

        # return id2word,   corpus, data_lemmatized
        return id2word, corpus, data_lemmatized ## Same as that perp returned var.

    # 保存并打印 LDA 模型的关键词及其权重值
    def save_and_print_top_keywords(lda_model, num_topics, num_words, filename):
        with open(filename, 'w') as file:
```

```python
        for topic_id in range(num_topics):
            words = lda_model.show_topic(topic_id, num_words)
            topic_keywords = ", ".join([f"{word} ({prob:.3f})" for word, prob in words])
            print(f"Topic {topic_id + 1}: {topic_keywords}")
            file.write(f"Topic {topic_id + 1}: {topic_keywords}\n")

# 可视化模型的困惑度并保存
def perplexity_visible_model(dictionary, corpus, texts, topic_num):
    x_list = []
    y_list = []
    with open("perp-val-f.txt", "w") as file:
        for i in range(1, topic_num):
            lda_model = LdaModel(corpus=corpus,
                                 id2word=dictionary,
                                 num_topics=i,
                                 random_state=100,
                                 chunksize=100,
                                 passes=10,
                                 alpha='auto',
                                 eta='auto')
            perplexity = lda_model.log_perplexity(corpus)
            print(f"Topic {i}, Perplexity: {perplexity}")
            file.write(f"Topic {i}, Perplexity: {perplexity}\n")
            x_list.append(i)
            y_list.append(perplexity)

    plt.xlabel('主题数量')
    plt.ylabel('困惑度')
    plt.title('主题数量与困惑度关系')
    plt.plot(x_list, y_list)
    plt.savefig("perplexity-f.png")
    plt.show()
    plt.close()

if __name__ == "__main__":
    id2word, corpus, data_lemmatized = get_texts_dict_corpus()
    lda_model = LdaModel(corpus=corpus,
                         id2word=id2word,
                         num_topics=20,
                         random_state=100,
```

```python
                chunksize=100,
                passes=10,
                alpha='auto',
                eta='auto')
save_and_print_top_keywords(lda_model, 20, 10, "20-topics-keywds-perpN.txt")
perplexity_visible_model(id2word, corpus, data_lemmatized, 21)
print("困惑度验证脚本执行完成。")


# 可视化模型的一致性得分并保存一致性值
def coherence_visible_model(dictionary, corpus, texts):
    # 一致性计算和绘图
    x_list = []
    y_list = []
    with open("coherence_val-f.txt", "w") as file:
        for i in range(1, 21):
            lda_model = LdaModel(corpus=corpus,
                                 id2word=dictionary,
                                 num_topics=i,
                                 random_state=100,
                                 chunksize=100,
                                 passes=10,
                                 alpha='auto',
                                 eta='auto')
            cv = CoherenceModel(model=lda_model, texts=texts, dictionary=dictionary, coherence='c_v')
            coherence_score = cv.get_coherence()
            x_list.append(i)
            y_list.append(coherence_score)

            # 写入文件并打印一致性值
            coherence_info = f"主题数: {i}, 一致性得分: {coherence_score}\n"
            file.write(coherence_info)
            print(coherence_info)

    # 一致性绘图
    plt.plot(x_list, y_list)
    plt.xlabel('主题数量')
    plt.ylabel('一致性值')
    plt.title('主题数量与一致性值关系')
    plt.savefig("coherence_f.png")
```

```
    plt.show()
    plt.close()

if __name__ == "__main__":
    dictionary, corpus, texts = get_texts_dict_corpus()
    coherence_visible_model(dictionary, corpus, texts)
    print("一致性验证脚本执行完成。")
```

运行以上代码，生成 LDA 主题与困惑度关系图，如图 7-3 所示；生成 LDA 主题与一致性关系图，如图 7-4 所示。图 7-4 显示主题为 20 个时，困惑度最低，但不能排除过拟合情况，因此需要结合主题与一致性关系图情况，图 7-4 显示主题为 20 个时，模型一致性最高，通常认为 20 个主题时模型稳定可靠，这种情况下给出的关键词可信度较高。

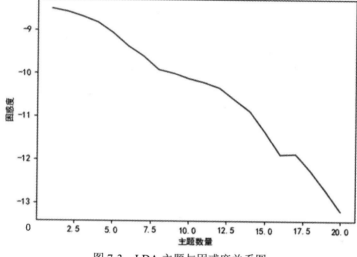

图 7-3　LDA 主题与困惑度关系图

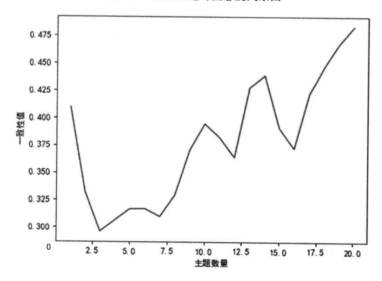

图 7-4　主题数与一致性关系图

在实际操作中，仅凭困惑度和一致性确定模型也不一定完全可靠，因此，需要同时参考模型给出的关键词推测主题，也需要调整模型的参数研究不同主题数量情况下的困惑度、一致性和关键词情况，最终主题数目需要综合考虑以上因素。最佳主题数目给出的关键词可信度相对较高。

7. GPT 模型

调用 Openai API 可以使用 GPT-3.5 系列模型，ChatGPT 订阅用户可以使用 GPT-4 系列模型。其中，GPT-3.5 系列中最有能力且使用费用较低的模型是 gpt-3.5-turbo，它已经针对使用 Chat Completions API 的聊天进行了优化，也适合完成传统任务。GPT-4 是一个大型多模态模型，接受文本或图像输入并输出文本，与 gpt-3.5-turbo 一样，GPT-4 也针对聊天进行了优化。

GPT-4 在性能上超越了之前的大型语言模型，它不仅在英语中的表现超过现有模型，而且在其他语言中也表现出强大的性能。最新的 GPT-4 模型具有改进的指令跟踪、JSON 模式、可复制的输出、并行函数调用等更多功能。gpt-4-1106-preview 是目前最新模型之一，窗口上下文长度可达 128k 形符，返回最多 4096 个输出形符。

调用 Openai API KEY，提取"imdb_review-s-100b.csv"文件"review"列文本中的前 50 个关键词并保存在名为"openaiKeywds.txt"文件中。代码如下：

```
import openai
import pandas as pd

# 加载 CSV 文件
df = pd.read_csv("imdb_review-s-100b.csv", encoding='Latin-1')

# 将'review'列中所有非字符串值转换为字符串
df['review'] = df['review'].fillna('').astype(str)

# 设置 OpenAI API 密钥
openai.api_key = 'YOUR-OPENAI-API-KEY'

# 使用 OpenAI 的 GPT-4 模型提取关键词的函数
def extract_keywords(text):
    try:
        response = openai.ChatCompletion.create(
            model="gpt-4-1106-preview",
            messages=[
                {"role": "system", "content": "You are a helpful assistant."},
                {"role": "user", "content": f"Extract 50 keywords from this text: {text}"}
            ]
        )
        # 提取关键词并去除前后空格
```

```
        keywords = response['choices'][0]['message']['content'].strip()
        print(f"Extracted Keywords: {keywords}")
        return keywords
    except Exception as e:
        # 打印 OpenAI 调用过程中的错误信息
        print(f"Error in OpenAI call: {e}")
        return None

# 将所有"review"列评论合并为一段文本
all_reviews = ' '.join(df['review'].tolist())

# 从合并后的文本中提取关键词
keywords = extract_keywords(all_reviews)

# 如果提取到关键词,将它们保存到文件中
if keywords:
    with open("openaiKeywds.txt", "w") as file:
        file.write(keywords)

print("Keyword extraction completed.")
```

运行以上代码,打印生成的关键词并保存至指定的文件中,生成的关键词数量超出指定要求。在使用同样指令的情况下,经过输入不同长度文本测试,发现使用较短文本时可以按照指令数量生成关键词,输入长文本后生成的关键词数量与所要求的字数不符,同时发现不论输入的文本长短,生成的关键词不仅包括单个单词,也包括 2 个及以上单词的关键词。

7.6 提取文本摘要

文本摘要生成有两种基本方法,分别是提取法和抽象法。提取法是从原始文本中提取单词和单词短语来创建摘要。抽象法是学习内部语言表示以生成更像人类的摘要,来解释原始文本的意图。

(1) 使用 HanLP 提取文本摘要。HanLP 的主要功能包括分词、词性标注、关键词提取、自动摘要、依存句法分析、命名实体识别、短语提取、拼音转换和简体繁体汉字转换等。代码示例如下:

```
from pyhanlp import *                    #自动下载 pyhanlp
text = open (r"filepath\filename.txt",encoding='gb18030',errors = 'ignore').read()
print(HanLP.extractKeyword(text, 5))     #自动提取 5 个关键词
print(HanLP.extractPhrase(text, 5))      #自动提取 5 个短语
```

```
print(HanLP.extractSummary(text, 10))    #自动提取 10 条文本摘要
```
代码中提取关键词、短语和摘要的数量可以设定。

使用 HanLP 提取文本摘要是一种针对中文优化的方法，它提供多种算法，如 TextRank 和 MMR，支持自定义词典，集成了多种 NLP 功能，使用方便。但它对非中文文本支持较弱，计算资源需求高，且需要一定的配置和调优。这种方法仅基于关键词和特征，可能无法捕捉文本的整体意义。

(2) 使用 SnowNLP 提取文本摘要。示例代码如下：

```
from snownlp import SnowNLP
text = open(r"filepath\filename.txt",encoding='gb18030',errors = 'ignore').read()
s = SnowNLP(text)
print(s.keywords(3))    #自动提取 3 个关键词
print(s.summary(3))    #自动提取 10 条文本摘要
```

SnowNLP 是轻量级的中文文本处理工具，易于安装和使用，除摘要外还提供情感分析等功能。但它的功能相对简单，摘要算法选择有限，对长文本或复杂文本的处理能力可能不足。与 HanLP 类似，SnowNLP 也是基于关键词的提取方法，可能忽略了文本的上下文。

(3) 使用 genism 模块及其汇总函数提取文本摘要。示例代码如下：

```
from gensim.summarization import summarize
text=open(r"filepath\filename.txt",encoding="gbk").read()
print(summarize(text))
```

使用 gensim 模块及其汇总函数提取文本摘要是一种支持多语言的方法，提供多种主题建模和文本相似度算法，社区活跃，文档丰富。它可以处理大规模文本数据，但主要面向英文，某些功能需要较深的机器学习知识，摘要质量很大程度依赖于预处理和参数调整。此外，Gensim 的训练和调优可能需要更多的时间和资源。

(4) TF-IDF，Pagerank 加载停用词提取文本摘要。示例代码如下：

```
import jieba
import networkx as ntx
from sklearn.feature_extraction.text import TfidfVectorizer, TfidfTransformer

def cut_sentence(sentence):
    delimiters = frozenset(u'。！？ ')
    buff = []
    for ch in sentence:
        buff.append(ch)
        if delimiters.__contains__(ch):
            yield ''.join(buff)
            buff = []
    if buff:
        yield ''.join(buff)
```

```python
def load_stopwords(path=r'filepath\filename.txt'):
    with open(path,encoding="gb18030") as file:
        # stopwords = filter(lambda x: x, list(map(lambda x: x.strip(), file.readlines())))
        stopwords = list(map(lambda x: x.strip(), file.readlines()))
    stopwords.extend([' ', '\t', '\n','.',',','。'])
    return frozenset(stopwords)
    return filter(lambda x: not stopwords.__contains__(x), jieba.cut(sentence))


def get_abstract(content, size=5):
    docs = list(cut_sentence(content))
    tfidf_model = TfidfVectorizer(tokenizer=jieba.cut, stop_words=load_stopwords())
    tfidf_matrix = tfidf_model.fit_transform(docs)
    normalized_matrix = TfidfTransformer().fit_transform(tfidf_matrix)
    similarity = ntx.from_scipy_sparse_matrix(normalized_matrix * normalized_matrix.T)
    scores = ntx.pagerank(similarity)
    tops = sorted(scores.items(), key=lambda x: x[1], reverse=True)

    size = min(size, len(docs))
    indices = list(map(lambda x: x[0], tops))[:size]
    return list(map(lambda idx: docs[idx], indices))


file1=open(r"filepath\filename.txt",encoding="gbk").read()
file1.close()
for i in get_abstract(text,5):
    print(i)
```

TF-IDF 和 PageRank 结合停用词的方法原理简单明了，计算效率高，适合处理大量文档且不依赖特定语言。但它仅基于统计特征，可能忽略语义信息，摘要质量受停用词表质量影响较大，对文本结构和上下文理解有限。因此，需要谨慎处理停用词。

（5）使用 GloVe 和 pagerank 提取文本摘要。GloVe(Global Vectors for Word Representation)是获取词向量的非监督学习算法，根据语料库的聚类词共现训练数据获得词向量。使用本地 GloVe 和 pagerank 提取文本摘要。示例代码如下：

```python
# 导入需要使用的库
import numpy as np
import pandas as pd
import nltk
nltk.download('punkt')
import re
#读取数据并分句
text=open(r"filepath\filename.txt",encoding="gbk").read()
```

```python
from nltk.tokenize import sent_tokenize
sentences = nltk.sent_tokenize(text)
#在斯坦福大学官网下载 GloVe 后保存在本地文件,调用 GloVe 词向量
word_embd_GloVe = {}
file = open(r'filepath\glove.6B.100d.txt', encoding='utf-8')
for line in file:
    values = line.split()
    word = values[0]
    coefs = np.asarray(values[1:], dtype='float32')
    word_embd_GloVe[word] = coefs
file.close()
# 清除标点符号、数字和特殊字符,仅保留英语
clean_sents = str(sentences).replace("[^a-zA-Z]", "")
# 全部字符变为小写
clean_sents = [s.lower() for s in clean_sents]
# 加载停用词
nltk.download('stopwords')
from nltk.corpus import stopwords
stop_words = stopwords.words('english')
# 使用自定义停用词表
stop_words=open(r'filepath\filename.txt').read()
# 去除停用词
def remove_stopwds(sen):
    sen_new = "".join([i for i in sen if i not in stop_words])
    return sen_new
# 去除 sentences 中所有停用词
clean_sents = [remove_stopwds(r.split()) for r in clean_sents]
# 用向量表示句子
sentence_vectors = []
for i in clean_sents:
    if len(i) != 0:
        vec = sum([word_embd_GloVe.get(w, np.zeros((100,))) for w in i.split()])/(len(i.split())+0.001)
    else:
        vec = np.zeros((100,))
    sentence_vectors.append(vec)
# 相似度矩阵
sim_mat = np.zeros([len(sentences), len(sentences)])
from sklearn.metrics.pairwise import cosine_similarity
```

```
for i in range(len(sentences)):
    for j in range(len(sentences)):
        if i != j:
            sim_mat[i][j] = cosine_similarity(sentence_vectors[i].reshape(1,100), sentence_vectors[j].reshape(1,100))[0,0]
# 应用pagerank
import networkx as nx
nx_graph = nx.from_numpy_array(sim_mat)
scores = nx.pagerank(nx_graph)
#提取摘要
ranked_sents = sorted(((scores[i],s) for i,s in enumerate(sentences)), reverse=True)
# 提取前10句作为摘要
for i in range(10):
    print(ranked_sents[i][1])
```

使用 GloVe 和 PageRank 提取文本摘要的方法结合了词向量的语义信息，可捕捉更深层次的文本特征，适用于多种语言。但这种方法需要预先训练或获取高质量的词向量，增加了复杂度，对计算资源需求较高，提取效果很大程度依赖词向量的质量和覆盖率。

选择合适的文本摘要方法时，需要根据具体应用场景、目标语言、文本规模等因素综合考虑。每种方法都有其优缺点，应根据实际需求权衡选择或结合使用。

7.7 词　　云

1. 英语词云

词云是文本关键词的可视化，它能快速提供文本的主要信息。目前有部分网站提供在线词云制作，Python 程序调用 WordCloud 模块也可生成词云。

【例 7-1】 加载停用词、调用本地文本和词云底图、设置最多显示 100 个词、随机数为 20 时生成英语文本的词云图。示例代码如下：

```
from wordcloud import WordCloud
import matplotlib
matplotlib.use("TkAgg")
import matplotlib.pylab as plt
import numpy
import PIL.Image as Image
stopwords={'will':0,'But':0,'the':0,'to':0,'and':0,'of':0,'for':0}
myfile=open('wcloudeng.txt','r',encoding='utf')
text=myfile.read()
print(text)
mask_pic=numpy.array(Image.open("词云底图2-1.png"))
```

```
wordcloud=WordCloud(background_color='orange',
    width=1000, height=800,
    margin=2, stopwords=stopwords,
    mask=mask_pic, max_words = 100, random_state=20).generate(text)
plt.imshow(wordcloud)
plt.axis('off')
plt.savefig("wordcld_en_fig.png")
plt.show()
```

运行以上程序,生成的词云如图 7-5 所示。

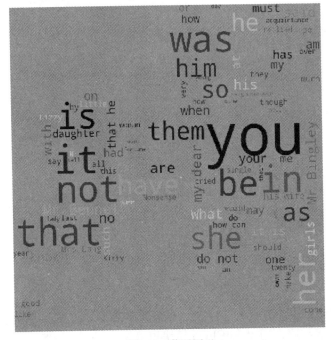

图 7-5　英语词云

2. 汉语词云

【例 7-2】 加载停用词、调用本地文本和词云底图、设置调用本地字体格式生成汉语文本的词云图。示例代码如下:

```
from wordcloud import WordCloud
import matplotlib
matplotlib.use("TkAgg")
import matplotlib.pylab as plt
import numpy
import PIL.Image as Image
stopwords={'我':0,'他':0}
myfile=open('词云-永别了武器.txt','r',encoding='utf-8')
text=myfile.read()
```

```
#print(text)
mask_pic=numpy.array(Image.open("词云底图 2-1.png"))
wordcloud=WordCloud(background_color='orange',
    width=1000,height=800,
    margin=2,stopwords=stopwords,
    mask=mask_pic,font_path="方正兰亭超细黑简体.TTF").generate(text)
image=wordcloud.to_image()
plt.savefig("wordcld_CN_fig.png")
image.show()
```

运行以上程序，生成的词云如图 7-6 所示。

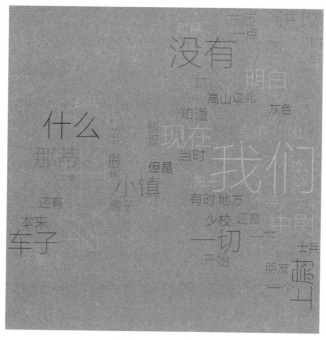

图 7-6　汉语词云

【例 7-3】　用未分词文本制作词云。示例代码如下：

```
from wordcloud import WordCloud
import matplotlib
matplotlib.use("TkAgg")
import matplotlib.pylab as plt
import jieba
import numpy
import PIL.Image as Image
stopwords={'美女':0,'情人':0}
myfile=open('JD_comments.txt','r',encoding='utf-8')
# 读取文本并分词
text=myfile.read()
```

```
split="".join(jieba.cut(text,cut_all=False))
print(split)


mask_pic=numpy.array(Image.open("词云底图1.jpg"))
wordcloud=WordCloud(background_color='orange',
    width=1000,height=800,
    margin=2,stopwords=stopwords,
    mask=mask_pic, max_words=100, font_path="FZLTXIHK.TTF").generate(split)
plt.imshow(wordcloud)
plt.axis('off')
plt.savefig('JD_wdcloud_fig.png')
plt.show()
```

运行以上程序，生成的词云如图7-7所示。

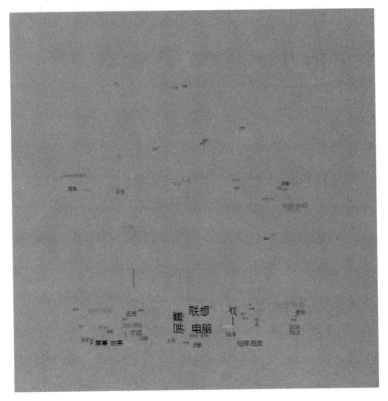

图7-7　京东商品评论词云图

使用WordCloud模块生成词云是一种直观、生动的文本可视化方法。这种方法的主要优点在于其简单易用性和高度可定制性。通过简单的几行代码，用户就能创建出美观的词云图像，直观地展示文本中的关键词及其重要性。WordCloud支持多种参数调整，如字体、颜色、形状等，使得用户可以根据需求创建独特的词云效果。此外，它还能与其他Python库(如matplotlib)无缝集成，方便进一步的图像处理和展示。

使用 WordCloud 也存在一些局限性。首先，词云主要关注单词频率，可能忽略词语间的语义关系和上下文信息。对于需要深入文本分析的应用场景，仅依赖词云可能不够全面。其次，对于包含大量词语的文本，生成高质量的词云需要较长的处理时间和较高的内存消耗。

在使用 WordCloud 时，需要注意以下事项。首先，需要注意文本的预处理，包括去除停用词、标点符号，以及进行词形还原等，这对生成有意义的词云至关重要。其次，合理设置词频阈值和最大词数可以帮助控制词云的复杂度和可读性。再次，选择合适的背景图片和颜色方案对提高词云的视觉效果和信息传达很有帮助。最后，对于中文文本还需要注意使用合适的分词工具(如 jieba)，以确保准确的词语切分。

总体上看，WordCloud 是一个强大而灵活的词云生成工具，适合用于文本摘要、主题展示等场景。但在使用时，需要根据具体需求和数据特点进行适当的配置和优化，以获得最佳的可视化效果。同时，也应该认识到词云的局限性，在需要深入文本分析的场景中，可以将其与其他文本分析方法结合使用，以获得对文本更全面的洞察。

第 8 章 情 感 分 析

8.1 情感分析原理

情感分析是对文本情感属性的挖掘，情感属性通常分为积极和消极两类，也可分为主观、客观和中性三类。使用情感分析可以判断海量文本的整体情感倾向，也可用来预测文本的情感倾向。情感分析可用来甄别事实和观点，通过分析评论文本，预测舆情及股票行情等。目前，可以使用第三方库实现基于词表的非监督方法情感分析，情感分析库包括 AFINN、MPQA、Senti WordNet、SnowNLP、TextBlob、BosonNLP、VADER 等，也可以通过标注数据、切分数据、表示词向量和训练模型等步骤来实现监督式情感分析。

非监督类情感分析方法主要是利用词性、词位置信息、词前后信息、短语信息、上下文信息等给词表内的词赋值。情感分析的最终结果是对文本中的词赋值的聚类结果，通过聚类结果可以分析文本情感、情态和主观性等。

以分析句子情感为例，首先需要判断句子中的情感词、程度词、感叹词、否定词、标点符号等表示情感的词或符号，然后给这些词和符号赋值后通过权重计算得出句子情感均值、积极情感值、消极情感值等。使用的算法或库不同，最终的情感指标不同，取值范围也不尽相同，如 Textblob 能给出句子的极性值和主观性值，SnowNLP 可能给出句子的情感均值。

8.2 TextBlob 情感分析

1. TextBlob 介绍

TextBlob 库可以用来提取名词短语、标记词性、分类文本、分句、分词、提取词干、计算词汇或短语频率、分析语法、获取 N 元序列、进行 WordNet 整合、在谷歌翻译支持下完成语言翻译和检查以及分析情感。

TextBlob 分析情感时返回的结果为元组形式。结果包括极性值，其值在 -1.0～1.0 之间，-1.0 表示消极，1.0 表示积极，值越大表示积极性越强；客观性结果值，其值在 0.0～1.0 之间，0.0 表示客观，1.0 表示主观，值越大表示主观性越强。

利用 TextBlob 分析单句情感。示例代码如下：

```
from textblob import TextBlob
test = TextBlob("Textblob is amazingly easy to use for beginners.")
print(test.sentiment)
# Sentiment(polarity=0.43333333333333335, subjectivity=0.8333333333333334)
```

```
print(test.sentiment.polarity)
# 0.43333333333333335
print(test.sentiment.subjectivity)
# 0.8333333333333334
```

2. TextBlob 文本情感分析及可视化

用 TextBlob 分析文本情感。示例代码如下:

```
from textblob import TextBlob
import pandas as pd
import pylab as pl
import numpy as np

txt = open(r'filepath\filename.txt', encoding='utf-8')
text = txt.readlines()
txt.close()
print('读入成功')

sentences = []
senti_score = []
for i in text:
    a1 = TextBlob(i)
    a2 = a1.sentiment
    sentences.append(i)
    senti_score.append(a2)
    print('doing')
table = pd.DataFrame(sentences, senti_score)
table.columns = ['sentences']   # 命名列
table.to_excel(r'filepath\filename.xlsx', sheet_name='Sheet1')
ts = pd.Series(sentences, senti_score)
ts = ts.cumsum()
print(table)
print(len(table))
print(type(table))

# 图示情感分析
import numpy as np
import matplotlib.pylab as plt
import matplotlib
matplotlib.use('TkAgg')
import matplotlib.pylab as plt
```

```python
plt.rcParams['font.sans-serif'] = ['SimHei']    # 用来正常显示中文标签
plt.rcParams['axes.unicode_minus'] = False    # 用来正常显示负号
import pandas as pd

df = pd.read_excel(r'filepath\filename.xlsx')
df.columns=['senti_values','sentences']   # 修改列名称

df[['number','polar_subj']] = df['senti_values'].str.split('\(|\)', expand=True).iloc[:,[0,1]]
num = df['number']
pol_sub = df['polar_subj']

df['polarity'], df['subjectivity'] = pol_sub.str.split(',', 1).str
polar = df['polarity']
subj = df['subjectivity']
print(polar)
print(subj)


x=range(0,len(table))
t_polar = pd.DataFrame(polar,dtype=np.float)   # 将 object 类型转为浮点形式
y1=t_polar
t_subj = pd.DataFrame(subj,dtype=np.float)
y2=t_subj
plt.plot(x,y1,label='polarity')
plt.plot(x,y2,label='subjectivity')
plt.title('CHNC-Sentiment by TextBlob',fontsize=8)   # 命名表名称
plt.xlabel(u'Abstracts Texts', fontsize = 8)
plt.ylabel(u'Sentiment Degrees', fontsize = 8)
polar_aver=np.mean(t_polar)
print(polar_aver)    # 0.08535157919111536
subj_aver=np.mean(t_subj)
print(subj_aver)    # 0.476675

plt.axhline(y=polar_aver.item(), color="red",linestyle = '--',label='polarity average')    # polarity 均值线
plt.axhline(y=subj_aver.item(), color="green",linestyle = '--',label='subjectivity average')    # subjectivity 均值线
plt.legend() # 图例
# plt.show()
plt.savefig(r"F:filepath\filename.jpg",dpi=300)
plt.ion() #显示图像
plt.pause(5) #显示 5 秒
```

plt.close() #关闭图像

TextBlob 情感分析结果如图 8-1 所示。

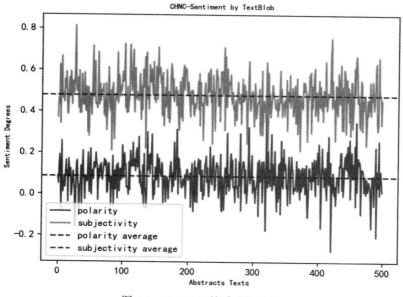

图 8-1　TextBlob 情感分析结果

TextBlob 情感分析功能基于预训练模型，可以快速给出文本的情感极性和主观性评分。TextBlob 的主要优点在使用简单、无需复杂的设置就能进行基本的情感分析。它还提供了其他 NLP 功能，如词性标注、名词短语提取等，使其成为一个多功能的文本处理工具。但 TextBlob 的情感分析模型主要是针对英文，对其他语言特别是中文的支持相对有限。此外，它的预训练模型可能不适合特定领域的文本，在处理专业术语或网络用语时可能表现欠佳。在使用 TextBlob 时，需要注意输入文本的语言和领域，必要时可能需要进行自定义训练以获取更好的情感分析结果。

8.3　SnowNLP 文本情感分析

1. SnowNLP 介绍

SnowNLP 是一个用 Python 写的类库，用来处理中文文本内容，它是受 TextBlob 的启发而写的。由于目前大部分自然语言处理库是针对英文的，因此科研人员编写了一个方便处理中文的类库。与 TextBlob 不同的是，SnowNLP 没有用 NLTK，所有的算法都是自己实现的，并且自带中文正负情感的训练集，使用朴素贝叶斯原理来训练和预测数据。

SnowNLP 可用来分词、标注词性和分析情感、标注拼音、提取关键字和摘要、计算词频和逆文档词频、转换繁体字为简体字、断句、计算文本相似度等。示例代码如下：

```
from snownlp import SnowNLP

s = SnowNLP(u'我这次网购的电脑很满意')
s1=SnowNLP(u'旅途很顺利，只是有点累；住处很安静，偶尔能听见鸟叫声')
```

```
print(s.words)
# ['我', '这次', '网', '购', '的', '电脑', '很', '满意']

for word, pos in s.tags:
    print(word, pos,end=' ')
# 我 r 这次 r 网 n 购 Vg 的 u 电脑 n 很 d 满意 v

p =s.sentiments
print(p)
# 0.5788596258982914

print(s1.sentiments)
# 0.9593298233976127
```

可以看出，第二句中尽管出现了个人感受"只是有点累"，但总体上是满意的，最终情感值得分为 0.959 329 823 397 612 7。

需要注意的是，SnowNLP 用于情感分析时训练数据主要是买卖东西时的评价，所以对其他类型文本的效果可能不是很好。它的分词能力没有 jieba 强大，如第一句分词中"网购"就被分成了"网"和"购"两个词，这显然是有问题的。

2. SnowNLP 情感分析

SnowNLP 情感分析源码主要由以下几个部分组成。

(1) 情感词典(sentimentdict.py)：该词典包含大约 8000 个常用的中文情感词语及其极性值，是情感分析的基础数据。

(2) Sentiment 类：位于 sentimentdict.py 文件中，负责实现基于情感词典的情感分析算法，其主要方法 sentiment(sentence)返回文本的情感极性值。

(3) naivebayes 包：提供了朴素贝叶斯分类器的实现，用于基于训练语料进行情感分类。该包的主要类包括 Classifier(朴素贝叶斯分类器主类)、BayesData(存储训练语料的数据结构)和一些数据预处理的辅助函数。sentiment.marshal 文件则是 SnowNLP 预先训练好的情感分类模型数据文件，使用朴素贝叶斯分类器在某一情感语料库上训练得到，提升了情感分析的准确性。

SnowNLP 类整合了分词、情感分析等多种功能，是主要的入口类。SnowNLP(sentence).sentiments 可以使用基于情感词典的方法进行情感分析，SnowNLP(sentence).sentiment Score 可以使用预训练的贝叶斯模型进行情感分类得分。此外，train.py 文件提供了在自定义语料上训练和持久化朴素贝叶斯分类器模型的功能，包括加载情感标注语料、初始化贝叶斯分类器、训练模型以及预测新文本的情感分数。

SnowNLP 情感分析首先将输入的文本按标点符号分割为多个短句，然后对每个短句进行分词和词性标注，接着对每个词进行情感极性值的查询和累加，通过计算短句中所有词的情感极性值，最终归为一个介于 −1～1 之间的浮点数来表示文本的情感极性。值越接近 1 表示情感越积极，越接近 −1 表示情感越消极。

SnowNLP 的情感分析模块结合了基于情感词典的方法和基于统计机器学习模型(朴素贝叶斯分类器)的方法, 为情感分析提供了两种不同的解决方案。用户可以根据具体需求和语料情况选择合适的分析方式, 从而在不同的应用场景中获得良好的效果。情感词典方法简单易用, 而贝叶斯分类器方法则依赖训练数据, 能在适当的语料支持下提供更高的准确性。

3. SnowNLP 文本情感分析及可视化

使用 SnowNLP 分析文本情感。示例代码如下:

```python
from snownlp import SnowNLP
import pandas as pd
import pylab as pl

import numpy as np

txt = open(r'CHTST-200.txt', encoding='gb18030')
text = txt.readlines()
txt.close()
print('读入成功')

sentences = []
senti_score = []
for i in text:
    a1 = SnowNLP(i)
    a2 = a1.sentiments
    sentences.append(i) #
    senti_score.append(a2)
    # print('doing')
table = pd.DataFrame(sentences, senti_score)
table.to_excel(r'Tstsnow.xlsx', sheet_name='Sheet1')

# print(table)
print(len(table))

x = range(0,len(sentences))
# print(x)
c=np.mean(senti_score) # 计算均值
print(c)

# 图示情感分析
import matplotlib.pylab as plt
```

```
import matplotlib
matplotlib.use('TkAgg')
import matplotlib.pylab as plt
plt.rcParams['font.sans-serif'] = ['SimHei']   # 用来正常显示中文标签
plt.rcParams['axes.unicode_minus'] = False   # 用来正常显示负号

plt.mpl.rcParams['font.sans-serif'] = ['SimHei']
plt.plot(x, senti_score)
plt.title(u'Figure N', fontsize = 8)   # u/U:表示 unicode 字符串
plt.xlabel(u'Abstract Texts', fontsize = 8)
plt.ylabel(u'Emotion Degrees', fontsize = 8)

plt.axhline(y=c, color="red",linestyle = '--')   # 均值线设置

plt.show()
plt.savefig('Tstsnowpic.png',dpi=300)
```

SnowNLP 情感分析结果如图 8-2 所示。

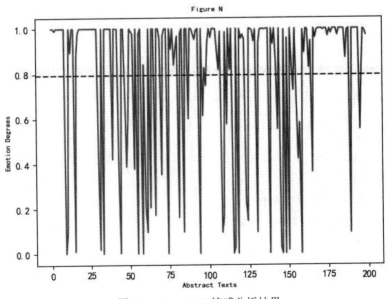

图 8-2　SnowNLP 情感分析结果

SnowNLP 情感分析功能针对中文文本进行了优化，能够较好地理解中文的语言意思和情感表达。SnowNLP 的优点包括对中文分词、词性标注的良好支持，以及相对准确的中文情感判断。它还提供了一些额外的功能，如文本摘要和关键词提取，这使得它成为处理中文文本的综合工具。但 SnowNLP 的模型相对简单，可能无法捕捉复杂的情感表达或处理含有大量网络用语的文本。另外，它的更新和维护频率不如一些更流行的 NLP 库。在使用 SnowNLP 时，应注意其应用范围主要针对中文文本，对其他语言的支持有限。同时，对于特定领域的文本，可能需要重新训练模型以提高准确性。

8.4 VADER 文本情感分析

1. VADER 介绍

VADER(Valence Aware Dictionary and sEntiment Reasoner)是一个基于词典和规则的情绪分析开源 Python 库，专门针对社交媒体中表达的情感进行调整，不需要使用文本数据进行训练，安装后就可以使用，适用于社交媒体等多种文本类型。VADER 文本情感分析能给出积极、消极、中性和综合分值，分值在 -1～1 之间。

2. VADER 的特点

VADER 分析情绪的特点。

(1) 使用标点符号可以增强情绪强度，而不改变情绪性质。例如，"The play is magic!"比"The play is magic!!!"更强烈。示例代码如下：

```
from vaderSentiment.vaderSentiment import SentimentIntensityAnalyzer
analyzer = SentimentIntensityAnalyzer()
vs1 = analyzer.polarity_scores('The play is magic!')
vs2 = analyzer.polarity_scores('The play is magic!!')
vs3 = analyzer.polarity_scores('The play is magic!!!')
print(vs1,'\n',vs2,'\n',vs3)
```

输出结果如下：

```
{'neg': 0.0, 'neu': 0.527, 'pos': 0.473, 'compound': 0.4003}
 {'neg': 0.0, 'neu': 0.501, 'pos': 0.499, 'compound': 0.4559}
 {'neg': 0.0, 'neu': 0.478, 'pos': 0.522, 'compound': 0.5067}
```

(2) 与情感相关的单词使用大写字母会增加情绪强度。例如"The library here is AWESOME!"传达的情感比"The library here is awesome!"要强。代码示例如下：

```
from vaderSentiment.vaderSentiment import SentimentIntensityAnalyzer
analyzer = SentimentIntensityAnalyzer()
vs4 = analyzer.polarity_scores('The library here is AWESOME!!')
vs5 = analyzer.polarity_scores('The library here is awesome!!')
print(vs4,'\n',vs5)
```

输出结果如下：

```
{'neg': 0.0, 'neu': 0.425, 'pos': 0.575, 'compound': 0.7519}
 {'neg': 0.0, 'neu': 0.461, 'pos': 0.539, 'compound': 0.6892}
```

(3) 通过一些语气词来增加或减少影响情绪强度。示例代码如下：

```
from vaderSentiment.vaderSentiment import SentimentIntensityAnalyzer
analyzer = SentimentIntensityAnalyzer()
vs6 = analyzer.polarity_scores('The book here is good!')
vs7 = analyzer.polarity_scores('The book here is quite good!')
vs8 = analyzer.polarity_scores('The book here is extremely good!')
```

```
print(vs6,'\n',vs7,'\n',vs8)
```

输出结果如下:

```
{'neg': 0.0, 'neu': 0.556, 'pos': 0.444, 'compound': 0.4926}
 {'neg': 0.0, 'neu': 0.589, 'pos': 0.411, 'compound': 0.54}
 {'neg': 0.0, 'neu': 0.589, 'pos': 0.411, 'compound': 0.54}
```

(4) 使用像"but"这样的连词表示情绪的变化。例如,"The bed is comfortable, but the service of the hotel is horrible"中的情绪喜忧参半,但转折词后的部分决定整体情绪评价。示例代码如下:

```
from vaderSentiment.vaderSentiment import SentimentIntensityAnalyzer
analyzer = SentimentIntensityAnalyzer()
vs9 = analyzer.polarity_scores('The bed is comfortable,but the service of the hotel is horrible')
    print(vs9)
{'neg': 0.259, 'neu': 0.741, 'pos': 0.0, 'compound': -0.5423}
```

(5) 句中的标点符号默认为中性,分值为1;不同表情符有不同的情感分值。示例代码如下:

```
from vaderSentiment.vaderSentiment import SentimentIntensityAnalyzer
analyzer = SentimentIntensityAnalyzer()
vs10 = analyzer.polarity_scores('she was ! yesterday')
vs11 = analyzer.polarity_scores('are you ?')
vs12 = analyzer.polarity_scores('☺')
print(vs10,'\n',vs11,'\n',vs12)
```

输出结果如下:

```
{'neg': 0.0, 'neu': 1.0, 'pos': 0.0, 'compound': 0.0}
 {'neg': 0.0, 'neu': 1.0, 'pos': 0.0, 'compound': 0.0}
 {'neg': 0.0, 'neu': 0.25, 'pos': 0.75, 'compound': 0.4588}
```

3. VADER 文本情感分析及可视化

使用 VADER 分析文本情感。示例代码如下:

```
# pip install vaderSentiment                # 安装 VADER 库
# pip install --upgrade vaderSentiment      # 升级 VADER 版本
from vaderSentiment.vaderSentiment import SentimentIntensityAnalyzer
# 读取文件
txt = open(r'CHNC-clean-para500.txt', encoding='utf-8')
# 按行读取
text = txt.readlines()
txt.close()
# print('读入成功')
# VADER 情感分析
analyzer = SentimentIntensityAnalyzer()
# 情感分析结果 字典数据转 CSV 文件
vs = [analyzer.polarity_scores(sentence) for sentence in text]
```

```python
# print(vs)                                    # 列表形式字典类 情感分析值
from pandas import DataFrame as DF
df = DF(vs)
df.to_csv("TstVader.csv")
# 图示情感分析
import numpy as np
import matplotlib.pylab as plt
import matplotlib
matplotlib.use('TkAgg')
import matplotlib.pylab as plt
plt.rcParams['font.sans-serif'] = ['SimHei']   # 用来正常显示中文标签
plt.rcParams['axes.unicode_minus'] = False     # 用来正常显示负号
import pandas as pd
#CSV 文件画图
df = pd.read_csv('TstVader.csv')
df.columns=['num','neg','neu','pos','compound']  # 修改列名称
# print(df)

neg = df['neg']
neu = df['neu']
pos = df['pos']
compound = df['compound']
print(len(neg))
# print(neg,neu,pos,compound)

x=range(0,len(neg))                             # neg 行数

t_neg = pd.DataFrame(neg,dtype=np.float)        # 将 object 类型转为浮点形式
y1=t_neg
t_neu = pd.DataFrame(neu,dtype=np.float)
y2=t_neu

y3 = pd.DataFrame(pos,dtype=np.float)
y4 = pd.DataFrame(compound,dtype=np.float)

plt.plot(x,y1,label='negative')
plt.plot(x,y2,label='neutral')
plt.plot(x,y3,label='positive')
plt.plot(x,y4,label='compound')
plt.title('Figure by VADER',fontsize=8)
```

```
plt.xlabel(u'Texts in CHNC', fontsize = 8)
plt.ylabel(u'Sentiment Degrees', fontsize = 8)
t_neg = pd.DataFrame(neg,dtype=np.float)    # 将 object 类型转为浮点形式
neg_aver=np.mean(t_neg)
# print(neg_aver)
t_neu = pd.DataFrame(neu,dtype=np.float)
neu_aver=np.mean(t_neu)
# print(neu_aver)

pos_aver = np.mean(y3)
# print(pos_aver)
compound_aver = np.mean(y4)
# print(compound_aver)
print(neg_aver, neu_aver, pos_aver, compound_aver)

plt.axhline(y=neg_aver.item(), color="red",linestyle = '--',label='negative average')    # neg 均值线
plt.axhline(y=neu_aver.item(), color="green",linestyle = '--',label='neutral average')   # neu 均值线
plt.axhline(y=pos_aver.item(), color="blue",linestyle = '--',label='positive average')   # pos 均值线
plt.axhline(y=compound_aver.item(), color="yellow",linestyle = '--',label='compound average')
                                                                        # compound 均值线

plt.legend()    # 图例
plt.savefig('TstVader.png',dpi=300)
plt.show()
```

执行程序，输出情感分析结果，如图 8-3 所示。

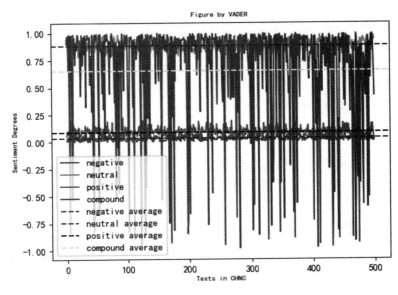

图 8-3　VADER 情感分析结果

VADER 是一个基于规则的情感分析工具，专门针对社交媒体文本设计。它的主要优势在于能够理解俚语、表情符号、缩写等社交媒体常见的表达方式，并考虑了标点符号和大小写等细节对情感的影响。VADER 不需要训练数据，可以直接应用于新的文本，这使得它在处理实时社交媒体数据时特别有效。它还能给出更细致的情感强度评分，而不仅仅是简单的正面或负面分类。但 VADER 主要是针对英语，对其他语言的支持有限。此外，由于它是基于规则的系统，在处理复杂或微妙的情感表达时可能不如基于机器学习的方法灵活。在使用 VADER 时，需要注意它最适合处理简短、非正式的英语文本，对于长篇文章或正式文档可能不太适用。

在选择情感分析方法时，需要考虑文本的语言、领域、长度以及分析的具体需求。对于英文文本，特别是社交媒体内容，VADER 可能是一个好选择。对于中文文本，SnowNLP 更为合适。TextBlob 则适合需要快速实现多种 NLP 功能的场景。无论选择哪种方法，都应该进行充分的测试和验证，以确保其在特定应用场景中的准确性。同时，对于高度专业化或特定领域的文本，需要考虑使用更高级的机器学习模型或进行自定义训练。

8.5 GPT 情感分析

GPT(Generative Pre-trained Transformer)是由美国 OpenAI 公司开发的大型语言模型。OpenAI 公司于 2018、2019 年分别发布了 GPT-1 和 GPT-2 两代模型，2020 年 6 月推出了第三代模型 GPT-3(2022 年 1 月统称最新第三代模型为 InstructGPT)，2022 年 3 月到 11 月相继开放了 text-davinci-002、code-davinci-002 和 text-davinci-003 等模型 API(这些模型和 gpt-3.5-turbo 等统称为 GPT-3.5)，同年 11 月 30 日发布了模型微调后的 ChatGPT，2023 年 3 月 14 日发布最新模型 GPT-4，4 月 10 日推出了 GPT-3.5 ALPHA 网页浏览版，随后发布了一系列 GPT 插件，GPT Plus 订阅用户可以使用 GPT-4 模型的相关功能。

GPT 在文本生成、理解、翻译、分类、推理，以及图像生成、代码编写等各个方面表现出色，是深度学习发展的新里程碑，必将对语言学习、研究等产生深远的影响。

情感分析可以帮助理解人们在社交媒体、产品评论、新闻文章等文本中表达的情感倾向。ChatGPT 利用循环神经网络(Recurrent Neural Network，RNN)等深度学习模型来进行情感分析。它根据输入的文本判断其是正面、负面还是中性的。ChatGPT 情感分析通过文本表示、序列建模、特征提取和情感预测等流程判断情感倾向，其基本原理是先将输入的文本转换为计算机可以处理的向量表示，然后使用 RNN 对文本序列进行建模，RNN 等模型根据词语的上下文信息、语法结构、重点词汇等特征对文本序列进行逐步处理，最后使用二元分类器输出层来预测文本的情感极性。

ChatGPT 使用 openai 库进行情感分析。例如，对于给定文本"我很喜欢电影《消失的她》，在电影院看这部影片时一直睡到片尾结束时才醒来！"分析情感，代码如下：

```
# 导入 openai 库
import openai
# 设置 OpenAI API 密钥，需要在 OpenAI 网站上申请这个密钥
openai.api_key = 'your-api-key'
```

```
# 创建一个聊天完成任务
# model 指定使用的模型, 这里使用的是"gpt-3.5-turbo"
# messages 是一个列表, 包含了一系列信息。每条信息是一个字典, 包含"role"和"content"两个键
# "role"可以是"system" "user"或者"assistant", 分别代表系统、用户和助手
# "system"通常用来设置场景, "user"和"assistant"则是聊天的内容
# 这里的场景是"You are a helpful assistant.", 意思是你是一个有益的助手
# 用户的信息是"Translate the following Chinese sentence to English and analyze its sentiment: '我很喜欢电影《消失的她》,在电影院看这部影片时一直睡到片尾结束时才醒来!'", 意思是翻译下面的中文句子并分析其情感: "我很喜欢电影《消失的她》,在电影院看这部影片时一直睡到片尾结束时才醒来!"
response = openai.ChatCompletion.create(
    model="gpt-3.5-turbo",
    messages=[
            {"role": "system", "content": "You are a helpful assistant."},
            {"role": "user", "content": "Translate the following Chinese sentence to English and analyze its sentiment: '我很喜欢电影《消失的她》,在电影院看这部影片时一直睡到片尾结束时才醒来!'"},
        ]
)
# 打印出聊天完成的结果
print(response['choices'][0]['message']['content'])
```

输出结果:

I really like the movie "The Disappeared Her". I fell asleep while watching this film in the cinema and woke up only at the end! Sentiment: The sentiment of the sentence is somewhat negative. Though the person expresses their liking for the movie, the fact that they fell asleep during the film suggests that they may not have found it engaging enough to hold their attention.

系统指示对输出结果有一定的影响, 如果把{"role": "system", "content": "You are a helpful assistant."}替换为{"role": "system", "content": "You are an experienced translator and sentiment expert."}, 输出结果如下:

I really like the movie "Disappeared Her". I fell asleep while watching this film in the cinema and woke up only at the end credits! Sentiment Analysis: The sentiment of this sentence is negative. The person expresses that they fell asleep during the movie and only woke up at the end. This implies that they found the movie boring or unengaging.

使用两种提示程序得到的结果不完全相同, 但总体结论基本一致。模型的性能可能会受到数据质量、文本长度、语言表达的多样性和程序指令等因素的影响, 尤其是对有隐含意义的文本情感分析仍需改进。此外, 情感分析的结果也具有一定的主观性, 不同的人可能对相同的文本有不同的情感理解。

第 9 章 机 器 翻 译

机器翻译(Machine Translation, MT)是指使用计算机自动将一种语言转换为另一种语言的过程，它是自然语言处理领域的一个重要研究方向。近年来，随着大规模预训练模型的改进，机器翻译的质量有了显著提高。

机器翻译的基本原理是通过训练模型，学习源语言和目标语言之间的对应关系，从而能够将源语言文本翻译成目标语言文本。其主要步骤和技术包括数据收集和预处理、特征提取和表示、模型选择和训练、解码和生成、质量评估和后处理等。常见的机器翻译有谷歌翻译、必应翻译、DeepL 翻译、百度翻译、有道翻译等；还有使用开源的预训练模型(如 Marian、GPT 模型以及 HuggingFace 社区的其他翻译模型等)的机器翻译。本章主要介绍 Marian、GPT、百度文心一言和 HuggingFace 社区的部分开源机器翻译。

9.1 Marian 机器翻译

Marian NMT(Neural Machine Translation)是由微软团队开发的一种基于 Transformer 的 seq2seq 模型，主要用于神经机器翻译任务；它是一种为研究人员和开发者提供高效和灵活机器翻译解决方案的框架。Marian 模型具有以下特点。

(1) 基于 Transformer 模型。Marian 主要使用基于 Transformer 的模型架构，这种架构在自然语言处理(NLP)领域已经广泛应用，并已经取得了卓越的性能。Transformer 模型使用自注意力机制，在翻译源语言文本到目标语言文本的过程中可以并行处理源文本的所有词，从而有效地处理了长距离依赖性，并且可以更快地进行训练。

(2) 高效的实现。Marian 框架对神经机器翻译进行了高效的实现，使其能够在各种硬件上进行高速运行，包括使用 CPU 或者 GPU。

(3) 多样的训练策略。Marian 支持多种训练策略，包括多 GPU 训练、模型平均等。

(4) 灵活的模型配置。Marian 允许用户通过配置文件灵活地定义模型的各种参数，这使得用户可以根据他们的具体需求和计算资源对模型进行调整。

(5) 集成的工具。Marian 提供了一些集成的工具，包括对数据进行预处理和后处理的工具，这些工具可以方便用户进行模型的训练和测试。

总的来说，Marian 是一种专门为神经机器翻译任务设计的强大工具，它的高效实现和灵活配置使其在机器翻译领域广受欢迎。

【例 9-1】 使用 MarianMT 模型把"谷歌翻译为全球用户提供了非常便捷的多语言翻译服务，推动了机器翻译技术的进步。"翻译成英语。代码如下：

```python
from transformers import MarianMTModel, MarianTokenizer
# 定义模型和分词器
model_name = 'Helsinki-NLP/opus-mt-zh-en'
tokenizer = MarianTokenizer.from_pretrained(model_name)
model = MarianMTModel.from_pretrained(model_name)
# 定义要翻译的文本
text = "谷歌翻译为全球用户提供了非常便捷的多语言翻译服务，推动了机器翻译技术的进步。"
# 对文本进行分词
tokenized_text = tokenizer.prepare_seq2seq_batch([text], return_tensors='pt')
# 翻译文本
translation = model.generate(**tokenized_text)
# 解码翻译结果
translated_text = tokenizer.decode(translation[0], skip_special_tokens=True)
# 打印翻译后的文本
print(translated_text)
```

运行代码，输出结果如下：

Google translation provides a very easy multilingual translation service for users around the world and contributes to technological advances in machine translation.

Marian 机器翻译非常高效，能够快速处理大量文本，适用于实时翻译场景。它支持多种语言和多种翻译模型，具有很强的灵活性和可扩展性。但它主要针对神经机器翻译，对于基于规则或统计的方法支持有限。此外，作为一个相对较新的系统，Marian 的用户社区和资源库可能不如一些更成熟的框架丰富。尽管 Marian 性能优异，但对于某些特定领域或低资源语言的翻译质量仍有提升空间。在处理复杂或具有多重含义的文本时容易出错，也存在无法理解语言习惯和文化差异的情况。

9.2 GPT 机器翻译

OpenAI 官网有多个模型，包括 text-davinci-003、GPT-3.5-turbo、GPT-4 等。目前，GPT-4 在各项任务上表现最好，但调用 GPT-4 API 需要申请。GPT-3.5 有多个版本，其中 GPT-3.5-turbo 版本经济且效果最佳，此外还有 GPT-3.5-turbo-16k、text-davinci-003、text-davinci-002、code-davinci-002 等模型。

1. 汉译英

【例 9-2】 调用 OpenAI API 把"机器翻译可以使用多种语言模型，其输出结果可能差异较大。" 翻译成英语。代码如下：

```
# 导入 openai 库，这是一个用于与 OpenAI API 交互的库
import openai
# 设置 OpenAI API 密钥。这是在 OpenAI 网站上注册后获得的一串字符，用于认证和调用 OpenAI 的服务
```

```
openai.api_key = "YOUR_API_KEY"
# 定义要翻译的文本
text = "机器翻译可以使用多种语言模型，其输出结果可能差异较大。"
# 调用 OpenAI 的 Completion 接口。该接口用于生成 AI 文本
# 需要指定模型名称(这里用的是 "text-davinci-003")，以及一系列的参数来控制文本生成的行为
response = openai.Completion.create(
    # 指定使用的模型名称
    engine="text-davinci-003",
    # 输入提示(prompt)给模型，提示内容是要求模型将给定的文本翻译成英文
    prompt=f"Translate this into English: {text}",
    # 温度参数决定了输出文本的随机性。值越大，输出结果的随机性越高；值越小，输出结果越确定
    temperature=0.3,
    # max_tokens 参数限制了生成文本的最大长度。这里设置的是 60 个词标记
    max_tokens=60,
    # top_p 参数是一个采样策略，又称为 nucleus sampling。设置为 1.0 表示将所有可能的下一个词标记都考虑进来
    top_p=1.0,
    # frequency_penalty 和 presence_penalty 参数用于影响模型生成的文本。这里都设置为 0，表示不对频率和存在进行惩罚
    frequency_penalty=0.0,
    presence_penalty=0.0
)
# 从响应中提取翻译的文本。这里的 'choices' 和 'text' 都是响应对象的属性，用于访问模型生成的文本
translation = response["choices"][0]["text"]

# 打印出翻译后的文本
print(translation)
```

运行代码，翻译结果如下：

Machine translation can use a variety of language models, and the output results may vary greatly.

【例 9-3】 使用"GPT-3.5-turbo"模型把"机器翻译可以使用多种语言模型，其输出结果可能差异较大。"翻译成英语。代码如下：

```
# 导入 openai 库，这是一个用于与 OpenAI API 交互的库
import openai
# 设置 OpenAI API 密钥。这是在 OpenAI 网站上注册后获得的一串字符，用于认证和调用 OpenAI 的服务
openai.api_key = "YOUR-API-KEY"
# 定义一个函数 translate_text，它接受一个文本字符串作为输入
def translate_text(text):
    # 调用 OpenAI 的 ChatCompletion 接口。该接口用于生成对话式的 AI 文本
```

```python
# 需要指定模型名称(这里用的是 "gpt-3.5-turbo")和消息列表
response = openai.ChatCompletion.create(
    model="gpt-3.5-turbo",
    messages=[
        # 这个列表中的每个元素都是一个消息，其中包含角色("system" 或 "user")和内容
        # "system" 角色的消息通常用于设置场景或者对话的上下文
        {"role": "system", "content": "You are a helpful assistant.Translate the sentence into English"},
        # "user" 角色的消息则是用户的输入。这里的内容就是我们要翻译的文本
        {"role": "user", "content": text}
    ]
)
# 函数返回的是这个接口的响应，其中包含了模型生成的文本
return response

# 调用 translate_text 函数，将一段中文文本翻译成英文
response = translate_text("机器翻译可以使用多种语言模型，其输出结果可能差异较大。")
# 打印出翻译后的文本。这里的 'choices' 和 'message' 和 'content' 都是响应对象的属性，用于访问模型生成的文本
print(response['choices'][0]['message']['content'])
```

运行代码，翻译结果如下：

Machine translation can use multiple language models, and the output results may vary significantly.

在代码中使用了 OpenAI 的 Chat API 来与 GPT-3.5 模型进行对话。translate_text 函数接受要翻译的文本作为输入，通过调用 openai.ChatCompletion.create()方法来与模型进行交互。在 openai.ChatCompletion.create()方法中，指定了使用的模型为"gpt-3.5-turbo"，然后传递了一个消息列表。该列表包含了一个系统消息和一个用户消息。系统消息告诉模型它是一个有用的助手，而用户消息包含了要翻译的文本。

由于 GPT-3.5 模型是以英文为基础的，只能将文本翻译成英文。如果想翻译成其他语言，需要将输入文本和输出结果都进行相应的语言转换。

2. 英译汉

若需把英文翻译为汉语，可在读取文档后调用 NLTK 分词，再调用"gpt-3.5-turbo"模型翻译。代码示例如下：

```python
# 导入所需的库
from nltk.tokenize import word_tokenize  # 使用 NLTK 进行英语分词
import openai  # OpenAI 的官方库，用于调用 GPT-3 模型
openai.api_key = "YOUR_API_KEY"  # 设置 OpenAI API 密钥
# 读取要翻译的英文文本
with open('intrans.txt', encoding='utf-8') as f:  # 以 utf-8 编码方式读取文件
    english_text = f.read()  # 读取文件内容
# 使用 word_tokenize 函数对英文文本进行分词
```

```
english_tokens = word_tokenize(english_text)
# 创建提示，提示包含要翻译的英文分词文本
# 将 tokens 列表转回为字符串，因为 OpenAI API 接受的是字符串格式的输入
prompt = f"Translate from English to Chinese: {' '.join(english_tokens)}"
# 调用 OpenAI API
try:
    # 调用 ChatCompletion API 创建聊天任务
    response = openai.ChatCompletion.create(
        model='gpt-3.5-turbo',  # 指定使用的模型
        messages=[
            {"role":"system","content":"You are an expert translator in EAP for 20 years."},  # 系统角色的消息，设置聊天的背景
            {"role": "user", "content": prompt}  # 用户角色的消息，包含要翻译的文本
        ],
        temperature=1.0,  # 设置生成文本的随机性
        max_tokens=3800  # 设置生成文本的最大长度
    )
    # 从返回的结果中提取翻译的中文文本
    chinese_text = response["choices"][0]["message"]["content"]
    # 打印出翻译的中文文本
    print(chinese_text)
except Exception as e:  # 如果在调用 API 过程中出现任何异常，获取异常并打印
    print('Error:', e)
```

3. ChatGPT speak 插件

ChatGPT 翻译插件 speak 是一个专门为语言学习和跨文化交流设计的工具，可以通过订阅 ChatGPT Plus 安装调用此插件，然后在网页界面输入提示进行翻译，也可以通过麦克风输入语音提示后翻译。

1) speak 插件的优点

(1) 多语言支持。该插件支持多种语言，可以帮助用户翻译、解释和学习各种语言。

(2) 对上下文敏感。它不仅能提供直接的翻译，还能考虑到用户提供的上下文，如语境、语气和场合，以提供更准确的答案。

(3) 交互性良好。用户可以与插件进行互动，提供更多的上下文信息，以获得更准确的答案。

(4) 综合能力强。除了翻译，它还可以解释特定外语短语的意义和用法，或者解释如何在特定情境中使用某种语言。

2) speak 插件的缺点

(1) 依赖外部资源。为了提供准确的答案，speak 插件可能需要依赖外部的语言数据库或资源，这可能会增加响应时间。

(2) 可能出现误解。虽然 speak 插件能提供准确答案，但在某些复杂的语境或文化背景下，它仍然可能会误解用户的意图。

(3) 无法替代真实的语言学习。虽然 speak 插件是一个很好的工具，但它不能完全替代真实的语言学习经验或与真人交流的经验。

ChatGPT 的 speak 插件是一个强大的工具，特别是对于希望在语言学习和跨文化交流中获得即时帮助的用户。但它仍然有其局限性，用户应该结合其他资源和方法来最大化其学习效果。

ChatGPT Plus 用户也可以自定义 GPT 角色和输出要求，与非自定义模式相比，定制化模式下 GPT-3.5、GPT-4 默认模式和 GPT-4 插件模式生成的翻译结果略有差异。

ChatGPT 具有强大的上下文理解能力，能够捕捉句子间的语义关联，从而提供更加连贯和自然的翻译。它能够处理各种文体和主题的文本，从日常对话到专业文献都能应对。此外，ChatGPT 的多语言能力非常出色，可以在众多语言之间进行翻译，而无需为每种语言对建立专门的模型。但它在某些专业领域或特定语言环境中的翻译准确度可能不如专门的翻译系统。另外，ChatGPT 的输出有时可能会出现"幻觉"或不一致的问题，特别是在处理长文本时。与传统的机器翻译系统相比，ChatGPT 的计算资源需求较高，可能不适合某些需要高效率处理大量文本的场景。

9.3 百度翻译服务

百度翻译服务主要包括百度机器翻译、百度翻译开放平台和百度文心千帆大模型翻译等，使用不同类型的翻译 API 产生的翻译结果可能有所不同，在指令相同的情况下，使用同一模型也可能产生不同的翻译结果。本章主要介绍如何调用不同类型的 API 进行翻译。

百度机器翻译是基于人工智能技术的在线机器翻译服务，它使用了深度学习和神经网络等先进技术，采用了端到端的神经网络模型，这意味着它能够直接将源语言的文本映射到目标语言，不需要人工进行中间步骤的特征提取或对齐。这种方法在某种程度上提高了翻译的准确性和流畅性。

百度机器翻译支持多种语言之间的翻译，包括但不限于中文、英文、法文、德文、西班牙文、日文、韩文等。用户可以通过百度翻译官网、手机应用或其他集成了该服务的平台来进行翻译操作。

1. 百度机器翻译

【例 9-4】 调用百度机器翻译 API 把"Machine translation can use multiple language models, and its output results may vary significantly."翻译成汉语，代码如下：

```
# 导入 requests 库，用于发送 HTTP 请求
import requests
# 导入 json 库，用于处理 JSON 数据
import json
# 定义百度 API 密钥(API_KEY)和秘密密钥(SECRET_KEY)
API_KEY = "YOUR_API_KEY"
```

```python
SECRET_KEY = "YOUR_SECRET_KEY"
# 定义 main 函数，此函数是程序的主入口
def main():
    # 定义 API 的 URL
    url = "https://aip.baidubce.com/rpc/2.0/mt/texttrans/v1?access_token=" + get_access_token()

    # 构建请求的 payload，包含要翻译的文本以及源语言和目标语言
    payload = json.dumps({
        "from": "en",
        "to": "zh",
        "q": "Machine translation can use multiple language models, and its output results may vary significantly."
    })
    # 定义 HTTP 请求头，发送 JSON 格式的数据，返回 JSON 格式的数据
    headers = {
        'Content-Type': 'application/json',
        'Accept': 'application/json'
    }

    # 使用 requests 库的 request 函数发送 POST 请求，并获取响应
    response = requests.request("POST", url, headers=headers, data=payload)

    # 打印响应的文本内容
    print(response.text)


# 定义 get_access_token 函数，用于获取访问令牌
def get_access_token():
    """
    使用 AK，SK 生成鉴权签名(Access Token)
    :return: access_token，或是 None(如果错误)
    """
    # 定义获取访问令牌的 URL
    url = "https://aip.baidubce.com/oauth/2.0/token"
    # 构建请求的参数，包括授权类型，API 密钥和秘密密钥
    params = {"grant_type": "client_credentials", "client_id": API_KEY, "client_secret": SECRET_KEY}
    # 发送 POST 请求，并从返回的 JSON 数据中获取访问令牌
    return str(requests.post(url, params=params).json().get("access_token"))


# 如果当前脚本是直接运行，而不是被导入，就调用 main 函数
if __name__ == '__main__':
```

main()

运行代码，输出结果如下：

{"result":{"from":"en","trans_result":[{"dst":"机器翻译可以使用多种语言模型，其输出结果可能差异很大。","src":"Machine translation can use multiple language models, and its output results may vary significantly."}],"to":"zh"},"log_id":1680961228582422762}

百度机器翻译对中文的支持非常出色，能够准确把握中文的语言特点和文化内涵，这使得它在与中文相关的翻译任务中表现突出。它集成了大量的专业词库和行业术语，能够为多个领域提供较为准确的翻译。虽然它在中文相关的翻译中表现出色，但在某些小语种或非常专业领域的翻译质量可能不如其他国际知名的翻译系统。在处理长段落、复杂句式、有丰富文化背景或习俗的语言和具有复杂语境的文本时，有时会出现语义理解偏差，导致翻译结果不够流畅或准确。

2. 百度翻译开发平台

百度翻译开发平台是百度推出的一项开发者服务，旨在为开发者和企业提供机器翻译相关的 API 接口，以便将百度强大的机器翻译技术集成到各种应用和产品中。通过使用百度翻译开发平台，开发者可以轻松实现多语言之间的实时翻译和文本处理功能。

百度翻译开发平台的主要特点和功能如下：

(1) 多语言支持：百度翻译开发平台支持多种语言之间的翻译，涵盖了全球主要的语言，包括中文、英文、法文、德文、西班牙文、日文、韩文等。

(2) 翻译 API：开发者可以通过调用翻译 API，将需要翻译的文本发送到百度服务器，然后获取翻译后的结果。这使得开发者可以在自己的应用或网站中实现实时多语言翻译功能。

(3) 文本识别与翻译：除了直接翻译文本，百度翻译开发平台还提供文本识别与翻译功能，可以将图片中的文字识别并翻译成指定语言。

(4) 语音识别与翻译：百度翻译开发平台支持语音识别与翻译，可以将语音输入转换成文本，并进行相应的翻译处理。

(5) 通用场景翻译：百度翻译开发平台提供针对特定场景的翻译接口，如旅游、餐饮、医疗等行业，针对不同领域的翻译需求进行了相应的优化。

(6) 定制化翻译服务：开发者可以根据自己的需求，定制化翻译服务，以适应特定的业务场景。

(7) 简易接入：百度翻译开发平台提供了简洁易用的 API 接口和文档，方便开发者快速接入和集成机器翻译功能。

(8) 付费与免费套餐：百度翻译开发平台提供了付费和免费的套餐供开发者选择，免费套餐有一定的使用限制，付费套餐则提供更高级别的服务和更大的使用额度。

开发者可以通过访问百度翻译开发平台的官方网站，注册开发者账号，获取 API 密钥，并查阅相关的文档和示例代码，以便在自己的应用中集成百度机器翻译的功能。

【例 9-5】调用开发平台 API 把 "Machine translation can use multiple language models, and its output results may vary significantly." 翻译成汉语。示例代码如下：

```
import requests    # 导入 requests 库，用于发送 HTTP 请求
import hashlib     # 导入 hashlib 库，用于生成 MD5 摘要
```

```python
import json    # 导入 json 库，用于处理 JSON 数据

API_KEY = "YOUR_KEY"    # 你的 API 密钥
SECRET_KEY = "YOUR_KEY"    # 你的密钥

Source_text = "Machine translation can use multiple language models, and its output results may vary significantly."
# 待翻译的文本

def main():
    # 构建请求 URL
    url = "https://fanyi-api.baidu.com/api/trans/vip/translate?appid=" + API_KEY + "&q=" + Source_text + "&from=en&to=zh&salt=1435660288&sign=" + get_sign()

    headers = {'Content-Type': 'application/json'}    # 设置请求头的 Content-Type 为 application/json

    response = requests.request("GET", url, headers=headers)    # 发送 GET 请求获取翻译结果

    # 使用 UTF-8 编码解码响应内容
    response_data = response.content.decode('utf-8')

    # 加载 JSON 响应数据
    response_json = json.loads(response_data)

    # 检查响应中是否包含'trans_result'键
    if 'trans_result' in response_json:
        # 从响应中提取翻译文本
        translated_text = response_json['trans_result'][0]['dst']
        # 打印翻译文本
        print(translated_text)
    else:
        print("Translation failed. Response:", response_json)

def get_sign():
    str1 = API_KEY + Source_text + '1435660288' + SECRET_KEY    # 构造待签名的字符串
    m = hashlib.md5()    # 创建 MD5 对象
    m.update(str1.encode('utf-8'))    # 使用 UTF-8 编码对字符串进行摘要计算
    return m.hexdigest()    # 返回摘要的十六进制表示

if __name__ == '__main__':
```

```
    main()
```

运行以上代码，输出结果为"机器翻译可以使用多种语言模型，其输出结果可能差异很大。"

3. 文心千帆大模型平台

文心千帆大模型平台是面向企业开发者的一站式大模型开发及服务运行平台，提供基于文心一言底层模型(Ernie Bot)的数据管理、自动化模型定制微调以及预测服务云端部署一站式大模型定制服务，并提供可快速调用的文心一言企业级服务 API，助力各行业的生成式 AI 应用需求落地。

【例 9-6】 调用文心一言底层模型(Ernie Bot)把"Machine translation can use multiple language models, and its output results may vary significantly."翻译成汉语，示例代码如下：

```python
# 导入 requests 库，它是一个 Python HTTP 库，可以发送 HTTP 请求
import requests
# 导入 json 库，它可以用来处理 json 数据
import json

# 定义 API 的 key
API_KEY = "YOUR-KEY"
# 定义 API 的密钥
SECRET_KEY = "YOUR-KEY"

# 定义主函数
def main():
    # 定义请求的 URL
    url = "https://aip.baidubce.com/rpc/2.0/ai_custom/v1/
wenxinworkshop/chat/completions?access_token=" + get_access_token()

    # 定义请求的数据
    payload = json.dumps({
    "messages": [
        {
            "role": "user",
            "content": "translate the English sentence into Chinese."
        },
        {
            "role": "assistant",
            "content": "translate the English sentence into Chinese, no more answer to the sentence."
        },
        {
            "role": "user",
```

```
            "content": "Machine translation can use multiple language models, and its output results may vary significantly."
        }
    ]
})

    # 定义请求的头，其中包含了内容类型
    headers = {'Content-Type': 'application/json'}

    # 发送 POST 请求，并获取响应
    response = requests.request("POST", url, headers=headers, data=payload)

    # 打印响应的文本
    print(response.text)

# 定义获取 access token 的函数
def get_access_token():
    """
    使用 AK，SK 生成鉴权签名(Access Token)
    :return: access_token，或是 None(如果错误)
    """
    # 定义请求的 URL
    url = "https://aip.baidubce.com/oauth/2.0/token"
    # 定义请求的参数，其中包含了授权类型，客户端 ID 和客户端密钥
    params = {"grant_type": "client_credentials", "client_id": API_KEY, "client_secret": SECRET_KEY}
    # 发送 POST 请求，并获取响应，然后从响应中获取 access token
    return str(requests.post(url, params=params).json().get("access_token"))

# 如果这个文件是主文件，那么运行主函数
if __name__ == '__main__':
    main()
```

运行以上代码，输出结果如下：

{"id":"as-1jei6m40ak","object":"chat.completion","created":1690107125,"result":"机器翻译可以使用多种语言模型，其输出结果可能会有很大的差异。","is_truncated":false,"need_clear_history":false,"usage":{"prompt_tokens":40,"completion_tokens":28,"total_tokens":68}}

　　文心千帆大语言模型在理解上下文和处理复杂语言结构方面表现出色，这使得它能够提供更加流畅和符合语境的翻译。它对中文的理解和生成能力非常强，特别适合于与中文相关的翻译任务。文心千帆具有多模态能力，可以结合图像、文本等多种信息进行翻译，这在某些特定场景下非常有用。它具有良好的可扩展性，可以通过增加训练数据、调整模型

结构等方式进一步优化性能。由于它是一个通用的大型语言模型，而非专门为翻译任务优化的系统，在某些高度专业化的领域可能不如专用的翻译引擎准确。另外，作为一个相对较新的系统，它还没有经过足够多的实际应用检验，在稳定性和一致性方面还有提升空间。和其他大型语言模型一样，文心千帆在翻译过程中也会产生"幻觉"，生成看似合理但实际上不准确的内容。

9.4 Meta 机器翻译

1. M2M100 翻译模型

M2M100(Many to Many)是 Facebook 开发的一种大规模多语言神经机器翻译模型。它基于 Transformer 编码器-解码器结构，通过 Attention 机制捕获语言的长程依赖关系，是目前效果最好的多语言翻译模型之一，有语言覆盖广、模型规模大、翻译质量高等优点。M2M100 模型在 HuggingFace 平台上开源，可直接使用其 API 实现翻译，非常适合于翻译产品和研究。其主要特点如下：

(1) 覆盖面广。M2M100 支持超过 100 种语言的翻译，几乎覆盖了所有主流语言。

(2) 模型规模大。M2M100 模型容量达 417 亿个参数，是目前公开的最大规模的多语言翻译模型。

(3) 训练数据丰富。M2M100 使用了大规模的高质量翻译语料进行预训练，包括各类域的平行语料库。

(4) 支持多种翻译方式。M2M100 支持一对一、多对一、一对多等多种翻译场景。

(5) 性能强悍。M2M100 在各类翻译任务上的性能明显超过其他模型，翻译质量更高。

(6) 推理速度快。M2M100 使用各种优化技术，使得推理速度很快，可实现实时翻译。

基于 M2M100 实现简单机器翻译的主要步骤如下：

(1) 加载 M2M100 的模型和分词器。

(2) 定义 translate 翻译函数。

① 输入参数，包括源语言文本、源语言、目标语言。

② 使用分词器对源文本进行操作，得到输入标识符。

③ 调用模型的生成方法进行解码预测，输入源语言标识符，强制设置目标语言句首标记标识符。

④ 使用分词器对生成的输出标识符进行解码，得到翻译后的文本。

(3) 异常处理。使用 try-except 捕获并打印错误信息。

(4) 主函数部分。主要包括以下内容：

① 输入示例中文文本。

② 调用 translate 函数实现中英翻译。

③ 打印翻译结果。

【例 9-7】 通过 Transformers 和 M2M100 实现简单易用的机器翻译。示例代码如下：

```
from transformers import AutoModelForSeq2SeqLM, M2M100Tokenizer
```

```python
# Load model and tokenizer once on startup
model = AutoModelForSeq2SeqLM.from_pretrained("facebook/m2m100_418M")
tokenizer = M2M100Tokenizer.from_pretrained("facebook/m2m100_418M")

# Function to translate
def translate(text, src_lang, tgt_lang):
    # encode the source language text
    encoded = tokenizer.encode(text, return_tensors="pt", src_lang=src_lang)
    # generate translation
    outputs = model.generate(encoded, forced_bos_token_id=tokenizer.get_lang_id(tgt_lang))
    # decode the translation
    decoded = tokenizer.decode(outputs[0], skip_special_tokens=True)
    return decoded

# Main function
if __name__ == "__main__":
    text = "机器翻译为全球用户提供了非常便捷的多语言翻译服务。"
    print(translate(text, 'zh', 'en'))
```

运行代码，翻译结果如下：

The machine translation provides a very convenient multilingual translation service to global users.

【例 9-8】 读取本地文件，调用 M2M100 模型把汉语翻译成英语和法语并存入本地文件，示例代码如下：

```python
import os
from transformers import M2M100ForConditionalGeneration, M2M100-Tokenizer

model = M2M100ForConditionalGeneration.from_pretrained("facebook/m2m100_418M")
tokenizer = M2M100Tokenizer.from_pretrained("facebook/m2m100_418M")

def translate(text, src_lang, tgt_lang):
    tokenizer.src_lang = src_lang
    encoded_text = tokenizer(text, padding=True, truncation=True, max_length=1024, return_tensors="pt")
    generated_tokens = model.generate(input_ids=encoded_text
        ["inpu t_ids"], attention_mask=encoded_text["attention_mask"],forced_bos_token_id=tokenizer.
        lang_code_to_id[tgt_lang])
    output_text = tokenizer.batch_decode(generated_tokens, skip_special_tokens=True)
    return output_text

# Read the Chinese text from a local file
input_file = 'input_chinese.txt'
```

```
with open(input_file, 'r', encoding='utf-8') as f:
    chinese_text = f.read()

# Translate the Chinese text to English and French
english_translation = translate(chinese_text, "zh", "en")[0]
french_translation = translate(chinese_text, "zh", "fr")[0]

# Save the translated texts to local files
with open('output_english.txt', 'w', encoding='utf-8') as f:
    f.write(english_translation)

with open('output_french.txt', 'w', encoding='utf-8') as f:
    f.write(french_translation)

print("Translations saved to output_english.txt and output_french.txt")
```

运行代码，翻译结果"english translation""french translation"存储在程序所在同一文件夹内。

2. M4T 翻译模型

Meta（前 Facebook）推出了一种名为多模态机器翻译(Multimodal Machine Translation，M4T)的新型翻译模型，它可翻译 101 种语言，包括 96 种语言的文本翻译和 35 种语言的语音翻译。M4T 不仅使用文本数据，还整合了图像和声音数据，以提高翻译准确性。这一多模态方法在处理具有视觉和听觉信息的复杂场景时具有显著优势，例如在翻译包含图像或视频的社交媒体内容、医学影像报告或多媒体教学材料时。M4T 在以下几个方面表现出显著的多模态翻译能力：

(1) 信息整合。M4T 能够有效地整合来自不同模态的信息，以提供更准确和丰富的翻译输出。

(2) 上下文理解。通过对多模态数据的综合分析，M4T 能够更准确地理解上下文，从而生成更符合实际情境的翻译。

(3) 实时处理。由于其高度优化的模型架构和算法，M4T 能够在需要实时翻译的多模态场景(如实时新闻报道或在线教学)中快速生成高质量的翻译。

(4) 多领域适应性。M4T 的多模态翻译能力使其在医疗、教育、媒体和旅游等多个领域中具有广泛的应用前景。

用户可以通过网页使用 M4T 翻译，也可以使用 Python 程序调用该模型进行文本-文本翻译。

【例 9-9】 使用 Python 程序调用 M4T 模型进行翻译，代码如下：

```
# pip install transformers

from transformers import AutoTokenizer, AutoModelForSeq2SeqLM
```

```python
model_name = "model_name_here"  # 替换为"facebook/seamless-m4t-large, facebook/SeamlessM4T-Large(2.3B),或 facebook/SeamlessM4T-Medium(1.2B)"
tokenizer = AutoTokenizer.from_pretrained(model_name)
model = AutoModelForSeq2SeqLM.from_pretrained(model_name)

def translate(text):
    inputs = tokenizer(text, return_tensors="pt", padding=True)
    outputs = model.generate(**inputs)
    translated_text = tokenizer.batch_decode(outputs, skip_special_tokens=True)[0]
    return translated_text

if __name__ == "__main__":
    text = "我正在测试 M4T 的文本翻译功能。"
    translated_text = translate(text)
    print(f"Original Text: {text}")
    print(f"Translated Text: {translated_text}")
```

M4T 翻译模型可以进行文本、语音翻译，包括语音-语音(S2ST)、语音-文本(S2TT)、文本-语音(T2ST)、文本-文本(T2TT)和自动语音识别(ASR)等多种翻译。可以先安装该模型后再实例化翻译对象，使用该模型完成多种翻译任务。

目前可以通过 Huggingface 网页的 seamless_communication 包获取 SeamlessM4T 模型，按照安装指导说明进行安装。

安装完成后，实例化一个 Translator 对象来执行所有五种口语任务。Translator 接受以下三个参数进行实例化。

(1) model_name_or_card：SeamlessM4T 检查点。可以是 seamlessM4T_medium，代表中等模型；或者 seamlessM4T_large，代表大型模型。

(2) vocoder_name_or_card：声码器检查点(vocoder_36langs)。

(3) device：Torch 设备。示例代码如下：

```
import torch
from seamless_communication.models.inference import Translator

# 使用多任务模型和声码器在 GPU 上初始化 Translator 对象
translator = Translator("seamlessM4T_large", vocoder_name_or_card="vocoder_36langs", device=torch.device("cuda:0"))
```

实例化后使用 predict()方法在任何支持的任务上多次运行推理。给定一个输入音频 <path_to_input_audio> 或输入文本 <input_text>，其源语言为 <src_lang>，我们可以按照以下方式翻译为目标语言 <tgt_lang>。

1) S2ST 和 T2ST 翻译

```
# S2ST
```

```
translated_text, wav, sr = translator.predict(<path_to_input_audio>, "s2st", <tgt_lang>)


# T2ST
translated_text, wav, sr = translator.predict(<input_text>, "t2st", <tgt_lang>, src_lang=<src_lang>)
```

注意：对于 T2ST 必须指定<src_lang>。生成的单元会被合成，输出音频文件将用以下方式保存：

```
wav, sr = translator.synthesize_speech(<speech_units>, <tgt_lang>)


# 保存翻译后的音频生成
torchaudio.save(
    <path_to_save_audio>,
    wav[0].cpu(),
    sample_rate=sr,
)
```

2) S2TT、T2TT 和 ASR 翻译

```
# S2TT
translated_text, _, _ = translator.predict(<path_to_input_audio>, "s2tt", <tgt_lang>)


# ASR
# 这与使用`<tgt_lang>=<src_lang>`的 S2TT 等效
transcribed_text, _, _ = translator.predict(<path_to_input_audio>, "asr", <src_lang>)


# T2TT
translated_text, _, _ = translator.predict(<input_text>, "t2tt", <tgt_lang>, src_lang=<src_lang>)
```

注意：对于 T2TT 必须指定<src_lang>。

Meta's M4T 在多模态机器翻译方面具有显著的优势和潜力，但也面临着多方面的挑战和问题，为未来的研究和应用提供了有价值的参考。特别是在人工智能和深度学习技术不断发展的今天，如何有效利用这些先进技术以提高多模态机器翻译的性能和应用，是一个值得进一步探讨的重要课题。

Meta 的机器翻译系统的主要优势在于其强大的数据基础和先进的神经网络模型。由于 Facebook 庞大的用户群和多语言内容，Meta 拥有大量的多语言数据，这使其翻译模型能够学习到丰富的语言表达和文化差异。此外，Meta 在开源方面做出了重要贡献，如发布了 FLORES-101 等多语言数据集，这推动了整个机器翻译领域的发展。Meta 的翻译系统还特别注重低资源语言的翻译能力，这对于提升全球语言的平等性具有重要意义。但作为一个主要服务于社交媒体平台的系统，在处理某些专业领域或正式文献的翻译时，Meta 可能不如专门的翻译系统准确。由于隐私问题，Meta 在使用用户数据进行模型训练时可能会受到一些限制，这可能会影响到某些语言或对特定领域的翻译质量。尽管 Meta 在开源方面做出了贡献，但其核心翻译技术和模型并未完全开放，这可能会限制外部研究者对其系统进行深入研究和改进。

9.5 Google 翻译

Google Translate 是 Google 的一个机器翻译系统，采用的是神经网络机器翻译技术。它通过大量的平行语料进行监督学习，可以提供多语言之间的自动翻译；利用深度神经网络，尤其是 seq2seq 和 Transformer 等模型，可实现端到端的神经机器翻译；采用无监督的神经网络技术不断优化翻译质量，并持续支持更多语言；支持超过 100 种语言之间的互译，覆盖了世界上绝大部分常用语言，为用户提供免费的翻译服务，可通过网页版、移动 APP、API 等不同形式使用；Google Translate 用户量很大，被广泛应用于语言学习、旅游等场景，也可用于语料采集和机器翻译研究。虽然近年来 Google Translate 的翻译质量持续提升，但仍存在一定程度的误译。

使用 Google Translate API 需要先在 GCP 控制台上启用该 API，并生成一个凭据文件下载到本地作为认证用。这种方式无须安装 googletrans 模块，直接调用 Google 提供的 Translate API 客户端即可。

1. Google Translate API 翻译

【例 9-10】 调用 Google Translate API 进行翻译。示例代码如下：

```python
import os
from google.cloud import translate_v2 as translate

os.environ['GOOGLE_APPLICATION_CREDENTIALS'] = '/path/to/credentials.json'

text = "谷歌翻译为全球用户提供了非常便捷的多语言翻译服务，推动了机器翻译技术的进步。"

client = translate.Client()
result = client.translate(text, target_language='en')

print(result['input'])
print(result['translatedText'])
```

运行代码，生成以下结果：

```
谷歌翻译为全球用户提供了非常便捷的多语言翻译服务，推动了机器翻译技术的进步。
Google Translate provides a very convenient multilingual translation service for users around the world, and promotes the advancement of machine translation technology.
```

Google Translate API 翻译的主要步骤如下。

(1) 设置环境变量，指定 Google Cloud Platform 凭据文件路径。
(2) 初始化 Translate 客户端。
(3) 调用 Translate 方法进行翻译，指定目标语言为英文。
(4) 输出原始文本和翻译结果。

【例 9-11】 翻译文档并存入本地磁盘。示例代码如下：

```python
import os
import httpx
from googletrans import Translator, LANGUAGES

# 设置谷歌云 API 密钥文件路径
os.environ["GOOGLE_APPLICATION_CREDENTIALS"] = " path/to/your/google_cloud_credentials.json"

# 读取本地翻译源语文件
input_file_path = " path/to/your/input_file.txt"
with open(input_file_path, "r", encoding="utf-8") as file:
    english_text = file.read()
#设置代理地址及端口
def httpx_translate(text, dest, src):
    proxies = {'http://': 'http://127.0.0.1:7890', 'https://': 'http://127.0.0.1:7890'}    # HTTP for both
    with httpx.Client(proxies=proxies, timeout=30.0) as client:
        response = client.get(
            "https://translate.googleapis.com/translate_a/single",
            params={
                "client": "gtx",
                "sl": src,
                "tl": dest,
                "dt": "t",
                "q": text,
            }
        )
        data = response.json()
        translated_text = data[0][0][0]
        return translated_text

# 翻译并保存译文文件

translated_text = httpx_translate(english_text, dest='zh-cn', src='en')

output_file_path = " path/to/your/output_file.txt"
with open(output_file_path, "w", encoding="utf-8") as file:
    file.write(translated_text)


print(translated_text)
print("Translation completed and saved.")
```

2. T5 机器翻译

T5(Text-to-Text Transfer Transformer)是由 Google 的研究团队于 2019 年开发的一种自然语言处理(NLP)模型。它的设计理念是将所有的 NLP 任务转化为一个文本到文本的转换任务,这种通用框架使得 T5 可以处理各种 NLP 任务,包括文本分类、问答、摘要生成、翻译等。

T5 是基于 Transformer 架构的,这种架构在 NLP 领域已经得到了广泛应用。Transformer 模型使用自注意力(Self-Attention)机制,该机制允许模型在处理每个词时考虑到整个文本的上下文,这使得 T5 能够更好地理解复杂和长距离的依赖关系。

在预训练阶段,T5 使用了一种被称为 Causal Language Modeling(CLM)或 Autoregressive 的预训练任务,它从左到右预测下一个词。在微调阶段,T5 将各种 NLP 任务都转化为"问题-答案"的形式。例如,在分类任务中,它会将一个分类任务转化为问题的形式,然后生成答案。

这种统一的框架使得 T5 在许多 NLP 基准测试中都取得了非常好的成绩。例如,在 GLUE、SuperGLUE 和 SQuAD 等基准测试中,T5 都表现优异。此外,T5 还提供了大小不同的模型版本,包括小型版本(T5-Small)、中型版本(T5-Base)、大型版本(T5-Large)、3B 版本(T5-3B)和 11B 版本(T5-11B),以满足不同计算资源和性能要求。总的来说,T5 是一种强大而灵活的 NLP 模型,适用于各种文本处理任务。

【例 9-12】 调用 "t5-base" 模型把 "Google translation provides a very easy multilingual translation service for users around the world and contributes to technological advances in machine translation." 翻译成法语。示例代码如下:

```
from transformers import pipeline

# 创建翻译 pipeline
translator = pipeline('translation_en_to_fr', model='t5-base')

# 定义英文文本
english_text = "Google translation provides a very easy multilingual translation service for users around the world and contributes to technological advances in machine translation."

#将文本翻译成法语
translation = translator(english_text, max_length=100)

# 打印翻译后的文本
print(translation[0]['translation_text'])
```

运行代码,输出结果如下:

Google translation offre un service de traduction multilingue très facile aux utilisateurs du monde entier et contribue aux progrès technologiques de la traduction automatique.

3. PaLM2 翻译

Google PaLM2 是谷歌于 2023 年 5 月 11 日发布的第二代大型语言模型(LLM)。它是在

PaLM 的基础上开发的，具有改进的编码、推理和多语言功能。PaLM2 基于谷歌的 Transformer 架构(该架构是一种用于自然语言处理的深度学习模型)，使用了大量数据来训练，包括超过 100 种的语言、科学数据集和代码。PaLM2 有四种不同大小的版本，分别命名为 Gecko、Otter、Bison 和 Unicorn。其中 Unicorn 是最大也是最强的版本，拥有超过 1000 亿个参数。

PaLM2 具有更好的编码能力和推理能力，可以更好地理解和翻译不同语言的文本，更好地理解不同语言之间的语义差异，从而提高翻译的准确性和流畅性。PaLM2 可以用于机器翻译、问答、自动生成文本和代码等任务，也可以用于开发新的人工智能算法、应用和教育等。

Google Vertex AI 是 Google Cloud 提供的一项人工智能服务，它提供了一系列功能来帮助开发者和企业构建、部署和管理人工智能模型。其中一项功能是翻译功能，它基于 Google 强大的神经网络机器翻译技术，能够提供高质量的翻译服务。Google Vertex AI 的翻译功能支持多种语言之间的翻译，包括汉语、英语、西班牙语、法语、德语、意大利语、葡萄牙语、俄语、日语、韩语等多种语言。它不仅支持文本翻译，还支持语音翻译和图像翻译。

Google Vertex AI 的翻译功能使用了最新的神经网络机器翻译技术，能够在保证翻译质量的同时提高翻译速度。此外，Google Vertex AI 的翻译功能还提供了多种 API 和 SDK，方便开发者和企业将其集成到自己的应用和服务中。它还提供了详细的文档和教程，帮助用户快速上手和使用。

使用 Google Vertex AI 调用 PaLM 机器翻译主要分为以下几个步骤。

(1) 启用谷歌云服务并创建 Vertex AI 项目。在谷歌云服务控制台启用 Vertex AI API 并创建一个 Vertex AI 项目，在控制台生成谷歌应用凭据。该项目将包含后续的翻译模型和资源。

(2) 上传训练数据。如果需要自定义模型，可以上传高质量的语料数据进行训练。数据要涵盖源语言和目标语言。

(3) 配置并训练自定义模型。在 Vertex AI 中配置自定义的神经机器翻译模型，选择参数和网络结构，然后进行训练。

(4) 部署翻译模型。训练完成后，可以直接部署模型提供在线翻译服务，也可以选择使用 Vertex AI 的预训练模型。

(5) 调用翻译 API。通过简单的 API 调用，传入源文本，获取翻译结果。该功能支持批量和单句翻译。

(6) 优化和更新。收集用户反馈，定期优化模型。可以通过增量训练来更新模型，提高翻译质量。

(7) 监控质量。Vertex AI 提供了丰富的监控指标，可以查看翻译质量、延迟等信息，并设置警报。

【例 9-13】 调用 chat-bison 模型翻译。示例代码如下：

```
# 导入所需的库和模块
import os
import vertexai
from vertexai.language_models import ChatModel, InputOutputTextPair

# 设置环境变量，指向 Google Cloud 凭证的 JSON 文件路径
```

```python
os.environ['GOOGLE_APPLICATION_CREDENTIALS'] = "path/to/your/google_cloud_credentials.json"

# 设置 HTTP 代理服务器的 IP 和端口(可选，根据需要进行设置)
os.environ['HTTP_PROXY'] = 'http://IP:PORT'

# 使用 Google Cloud 项目 ID 和位置初始化 Vertex AI
vertexai.init(project="Project ID", location="Your_location")

# 从 Vertex AI 加载预训练的聊天模型
chat_model = ChatModel.from_pretrained("chat-bison")

# 定义模型参数，这些参数将影响模型的输出
parameters = {
    "candidate_count": 1,        # 输出候选句子的数量
    "max_output_tokens": 1024,   # 最大输出令牌数(即输出的最大长度)
    "temperature": 0.2,          # 控制输出的随机性，值越高输出越多样，值越低输出越确定
    "top_p": 0.8,                # 控制输出的多样性，值越高输出越多样，但可能与主题相关性较低
    "top_k": 40                  # 控制输出多样性的另一种方式，值越高考虑的输出候选句子越多
}

# 开始一个聊天会话，并为模型提供上下文
chat = chat_model.start_chat(
    context="""You are an expert translator.""",
)

try:
    # 发送要翻译的文本消息给模型，并获取响应
    response = chat.send_message("""Translate the sentence into French:Content processed through Vertex AI is assessed against a list of safety attributes.""")
    print(f"Response from Model: {response.text}")   # 打印模型的响应文本
except google.api_core.exceptions.RetryError as e:   # 如果遇到重试错误,捕获该错误并打印相关消息
    print("RetryError encountered:", e)
```

运行以上代码，输出结果如下：

Response from Model:　Le contenu traité via Vertex AI est évalué par rapport à une liste d'attributs de sécurité.

【例 9-14】 调用 text-bison 模型翻译。示例代码如下：

```python
# 导入所需的库和模块
import os
import vertexai
from vertexai.language_models import TextGenerationModel
```

```python
# 设置环境变量，指向 Google Cloud 凭证的 JSON 文件路径
os.environ['GOOGLE_APPLICATION_CREDENTIALS'] = "path/to/your/google_cloud_credentials.json"

# 设置 HTTP 代理服务器的 IP 和端口(可选，根据需要进行设置)
os.environ['HTTP_PROXY'] = 'http://IP:PORT'

# 使用 Google Cloud 项目 ID 和位置初始化 Vertex AI
vertexai.init(project="gothic-isotope-400812", location="us-central1")

# 定义模型参数，这些参数将影响模型的输出
parameters = {
    "candidate_count": 1,        # 输出候选句子的数量
    "max_output_tokens": 1024,   # 最大输出令牌数(即输出的最大长度)
    "temperature": 0.2,          # 控制输出的随机性，值越高输出越多样，值越低输出越确定
    "top_p": 0.8,                # 控制输出的多样性，值越高输出越多样，但可能与主题相关性较低
    "top_k": 40                  # 控制输出多样性的另一种方式，值越高考虑的输出候选句子越多
}

# 从 Vertex AI 加载预训练的文本生成模型
model = TextGenerationModel.from_pretrained("text-bison")

# 使用模型进行预测，翻译给定的文本为中文
response = model.predict(
"""Translate the following into Chinese: Vertex AI Machine Translation is a fully managed service that provides fast, accurate, and affordable translation between languages. It is powered by Google's state-of-the-art neural machine translation technology, which has been shown to outperform traditional statistical machine translation methods. """ )

# 打印模型的响应文本
    print(f"Response from text-bison: {response.text}")
```

运行以上代码，输出以下结果：

Response from text-bison:
　　Vertex AI 机器翻译是一项完全托管的服务，可提供快速、准确进行翻译且价格合理。它由 Google 最先进的神经机器翻译技术提供支持，该技术已被证明优于传统的统计机器翻译。

4. BARD 翻译

　　Google BARD 是 Google AI 开发的基于 Transformer 架构的大型语言模型, 该架构是一种用于自然语言处理的深度学习模型。BARD 在大量的文本和代码数据集上进行了训练，能够生成逼真且信息丰富的文本，还可以翻译语言、编写不同类型的创意内容，并以丰富的

信息回答问题。

BARD 可以生成包括诗歌、代码、脚本、音乐作品、电子邮件、信件等各种类型的文本，可以翻译多种语言，包括汉语、英语、法语、西班牙语、德语等。

目前，有网页界面的 BARD 和非官方开发的 BARD API 可供使用。

【例 9-15】 调用 BARD API 模型翻译。代码示例如下：

```
# 导入所需的库和模块
# import os
from bardapi import Bard

# 设置 HTTP 代理服务器的 IP 和端口(可选，根据需要进行设置)
# Change 'http://proxy.example.com:8080' to your http proxy
proxies = {
    'http': 'http://IP:PORT',
    'https': 'https://IP:PORT'
}

# 初始化 Bard 对象，传入所需的参数(token、proxies 和 timeout)
bard = Bard(token='xxx.', proxies=proxies, timeout=60)

# 使用 Bard 对象获取问题的回答
# res = bard.get_answer("Do you like cookies?")['content']
# print(res)

# 使用 Bard 对象翻译给定的句子为中文、日文、法文和西班牙文
res = bard.get_answer("translate the sentence into Chinese, Japanese, French and Spanish:Human reviewers may process your Bard conversations for quality purposes. Don't enter sensitive info.")['content']
    print(res)
```

运行以上代码，输出以下结果：

Here is the translation of the sentence "Human reviewers may process your Bard conversations for quality purposes. Don't enter sensitive info." into Chinese, Japanese, French, and Spanish:

Chinese: 人工审核人员可能会出于质量目的处理您的 Bard 对话。请勿输入敏感信息。

Japanese: 人間の審査員は、品質向上のため、Bard での会話内容を処理することがあります。機密情報を 입력しないでください。

French: Des examinateurs humains peuvent traiter vos conversations Bard à des fins de qualité. N'entrez pas d'informations sensibles.

Spanish: Los revisores humanos pueden procesar tus conversaciones con Bard con fines de calidad. No introduzcas información sensible.

Please note that these are just machine translations, and it is always best to have a human translator review them if possible.

Google 翻译覆盖了世界上绝大多数常用语言，这使它成了一个真正的全球性翻译工具。它可以处理大量文本，速度快，能够在短时间内完成翻译任务。由于 Google 拥有海量的网络数据和用户反馈，其翻译质量在不断提升，特别是在常见语言对之间的翻译表现优秀。此外，Google 翻译提供了多样化的功能，如网页翻译、语音翻译、图片翻译等，大大增强了其实用性。Google 还开放了翻译 API，使开发者能够将翻译功能集成到各种应用中。尽管 Google 机器翻译总体上表现优秀，但在处理某些特定领域的专业术语或复杂的语言结构时，可能会出现错误或不准确的情况。对于一些低资源语言或罕见的语言对，翻译质量也可能不如专业的翻译工具。Google 翻译有时会过度依赖统计数据，导致某些翻译结果虽然常见但并不符合具体语境。

9.6 DeepL 翻译系统

DeepL 翻译系统是德国的 DeepL 公司于 2017 年推出由人工智能驱动的在线翻译工具，它的翻译引擎训练数据规模巨大，主要源自欧盟数据及其自身的数据集，基于先进的神经网络和深度学习技术，提供高质量的机器翻译服务。DeepL 支持多种语言，包括汉语、英语、德语、法语、西班牙语、意大利语、荷兰语、波兰语、葡萄牙语、俄语、日语等。

DeepL 致力于提供自然、准确的翻译结果，尽可能保持原文的风格、语义信息和情感。它能更好地理解和翻译句子的整体语境和语义，整体精确度很高、更加流畅、准确，中文翻译尤其有优势。

DeepL 也提供了开发者 API，允许开发者在自己的应用程序或网站中集成 DeepL 翻译。DeepL 翻译主要步骤如下：

(1) 导入 DeepL 模块并创建 Translator 对象，传入 DeepL 的授权密钥。
(2) 指定源文本和目标语言为英语(EN)。
(3) 调用 translate_text 方法进行翻译。
(4) 打印结果，result 包含原文和译文。

使用 DeepL API 需要先注册获取授权密钥。相比 Google Translate，DeepL 对与中文相关的翻译质量可能会更好，但语言数量较少。代码实现过程类似，主要是获取授权调用翻译接口。

【例 9-16】 调用 DeepL 翻译 API，把中文"谷歌翻译为全球用户提供了非常便捷的多语言翻译服务，推动了机器翻译技术的进步。"翻译成英文。示例代码如下：

```
import deepl
auth_key = 'YOUR_AUTH_KEY'
translator = deepl.Translator(auth_key)
text = "谷歌翻译为全球用户提供了非常便捷的多语言翻译服务，推动了机器翻译技术的进步。"
result = translator.translate_text(text, target_lang='EN-US')
print(result)
```

运行代码，翻译结果如下：

Google Translate provides global users with a very convenient multi-language translation service, and promotes the progress of machine translation technology.

注意：虽然 DeepL 的翻译质量在很多情况下都很出色，但和所有的机器翻译工具一样，它不可能完全取代人类的翻译，特别是在需要理解复杂语境、文化背景或细微情感的情况下。

【例 9-17】 使用 requests 库调用 DeepL 翻译 API。示例代码如下：

```python
# 导入所需的库
import requests    # 用于发送 HTTP 请求
import json    # 用于处理 JSON 格式的数据
# 定义一个函数，该函数接收要翻译的文本、目标语言和 DeepL API 密钥作为参数
def translate_text(text, target_language, api_key):
    # DeepL API 的 URL
    url = "https://api-free.deepl.com/v2/translate"
    # 要发送给 API 的数据
    data = {
        "auth_key": api_key,    # API 密钥
        "text": text,    # 要翻译的文本
        "target_lang": target_language,    # 目标语言
    }
    # 使用 requests 库的 post 方法发送请求，并将响应保存在 response 变量中
    response = requests.post(url, data=data)
    # 检查响应状态码，如果不是 200(表示请求成功)，则抛出异常
    if response.status_code != 200:
        raise Exception("DeepL API request failed with status code: {}".format(response.status_code))
    # 从响应中解析出翻译后的文本
    translated_text = json.loads(response.text)['translations'][0]['text']

    # 返回翻译后的文本
    return translated_text
# 要翻译的文本
text = "谷歌翻译为全球用户提供了非常便捷的多语言翻译服务，推动了机器翻译技术的进步。"
# 你的 DeepL API 密钥(请替换为你的实际密钥)
api_key = "your_api_key"
# 目标语言(这里是英语)
target_language = "EN"
# 调用 translate_text 函数进行翻译，并将结果保存在 translated_text 变量中
translated_text = translate_text(text, target_language, api_key)
# 打印翻译后的文本
print(translated_text)
```

运行代码，翻译结果如下：

Google Translate provides global users with a very convenient multi-language translation service, and promotes the progress of machine translation technology.

使用代码时需要替换 your_api_key 为实际 DeepL API 密钥。【例 9-17】的代码是将文本翻译成英文。如果想翻译成其他语言，可以更改 target_language 变量的值。例如，DE 表示德语，FR 表示法语等。

DeepL 翻译系统在处理欧洲语言时表现突出。它能够准确把握语言的细微差别和上下文含义，翻译结果通常更加流畅自然。在处理专业术语和长句复杂结构时表现优异，这使它在学术和商业领域得到广泛应用。DeepL 提供了网页界面和 API 接口，便于集成到各种工作流程中。虽然 DeepL 在欧洲语言的翻译质量上表现出色，但它支持的语言数量相对较少，特别是在亚洲语言和小语种方面的覆盖不如谷歌等大型竞争对手。作为一个商业服务，DeepL 的一些高级功能和大量使用需要付费，这可能会限制一些用户的使用。另外，尽管整体质量很高，但 DeepL 有时也会在某些特定领域或罕见表达的翻译中出现错误。由于 DeepL 的技术细节较为封闭，研究人员难以深入了解和改进其底层算法。

9.7　讯飞星火认知模型翻译

讯飞星火认知大模型是科大讯飞股份有限公司研发的以中文为核心的新一代认知智能大模型。它在与人进行自然对话互动过程中，同时提供以下多种能力：

(1) 内容生成能力：可以生成多风格、多任务的长文本，如邮件、文案、公文、作文、对话等。

(2) 语言理解能力：可以进行多层次、跨语种的语言理解，实现语法检查、要素抽取、语篇归整、文本摘要、情感分析、多语言翻译等功能。

(3) 知识问答能力：可以回答各种各样的问题，包括生活知识、工作技能、医学知识等。

(4) 推理能力：拥有基于思维链的推理能力，能够进行科学推理、常识推理等。

(5) 多题型步骤级数学能力：具备数学思维，能理解数学问题，覆盖多种题型，并能给出解题步骤。

(6) 代码理解与生成能力：可以进行代码理解、代码修改以及代码生成等工作。

讯飞星火认知大模型提供了对话式 AI 服务能力与解决方案，包括多轮对话、知识问答、逻辑推理、多模态理解、文本生成、语言生成等，并建立了基于思维链的开放式知识问答系统的网页用户界面。同时，为了方便开发者快速使用星火认知大模型，提供了简单易用的 API 接口和 SDK 支持，帮助开发者快速集成讯飞星火认知大模型功能到自己的应用中。讯飞星火认知大模型目前已经全面开放 API 接口。可访问其官方网站了解讯飞星火 API 的详细内容，并直接免费领取试用 tokens。通过讯飞开放平台提供的 REST API 接口，可以直接调用星火认知大模型的服务和功能，进行开发和部署。同时还提供了详细的文档和示例代码，帮助用户快速上手和使用。

调用星火 API 翻译时，需要进入讯飞开放平台快捷登录页，通过微信扫码、手机登录，即可快速成为讯飞开放平台的注册开发者；或进入讯飞开放平台注册页，注册完整的开放平台账号成为讯飞开放平台注册开发者。登录平台后，通过右上角"控制台"，或右上角下拉菜单的"我的应用"进入控制台。若未曾创建过应用，按照引导创建第一个应用。完成实名认证，进入"我的应用"，在应用中选择"机器翻译"或者"机器翻译 niutrans"，领取免费套餐或购买付费套餐。用于机器翻译的服务接口认证信息 APPID、APISecret、APIKey 三个参数和请求地址位于页面右侧。

基于讯飞自主研发的机器翻译引擎，已经支持包括中、英、日、法、西、俄等 70 多种语言。通过调用 API 接口，可将源语种文字转化为目标语种文字。在讯飞开放平台机器翻译 API 文档接口说明中有 Python 等语言的机器翻译 API 调用说明和示例代码。

【例 9-18】 调用讯飞星火模型 API 翻译系统，将英文 "Machine translation has made great progress in recent years, but there is still a lot of room for improvement." 翻译成中文。示例代码如下：

```python
# 导入所需库
from datetime import datetime
from wsgiref.handlers import format_date_time
from time import mktime
import hashlib
import base64
import hmac
from urllib.parse import urlencode
import json
import requests

# 配置你的讯飞开放平台的 APPId、APISecret 和 APIKey
# APPId = "YOUR_APPId"       #填写控制台中获取的 APPID 信息
# APISecret = "YOUR_APISecret"   #填写控制台中获取的 APISecret 信息
# APIKey = "YOUR_APIKey"     #填写控制台中获取的 APIKey 信息

# 配置术语资源唯一标识，如果不使用术语资源可以不设置这个参数
RES_ID = "its_en_cn_word"

# 设置需要翻译的原文本内容
TEXT = "Machine translation has made great progress in recent years, but there is still a lot of room for improvement."

# 自定义异常类，用于处理 assemble header 时的错误
class AssembleHeaderException(Exception):
    def __init__(self, msg):
        self.message = msg

# Url 类，用于存储解析后的 URL 组件
class Url:
    def __init__(self, host, path, schema):
```

```python
        self.host = host
        self.path = path
        self.schema = schema

# 计算给定数据的 SHA-256 哈希值并将其编码为 Base64 字符串
def sha256base64(data):
    sha256 = hashlib.sha256()
    sha256.update(data)
    digest = base64.b64encode(sha256.digest()).decode(encoding='utf-8')
    return digest

# 解析给定的 URL，返回一个包含 host、path 和 schema 的 Url 对象
def parse_url(requset_url):
    stidx = requset_url.index("://")
    host = requset_url[stidx + 3:]
    schema = requset_url[:stidx + 3]
    edidx = host.index("/")
    if edidx <= 0:
        raise AssembleHeaderException("invalid request url:" + requset_url)
    path = host[edidx:]
    host = host[:edidx]
    u = Url(host, path, schema)
    return u

# 构建 WebSocket 鉴权请求 URL
def assemble_ws_auth_url(requset_url, method="POST", api_key="", api_secret=""):
    u = parse_url(requset_url)
    host = u.host
    path = u.path
    now = datetime.now()
    date = format_date_time(mktime(now.timetuple()))
    signature_origin = "host: {}\ndate: {}\n{} {} HTTP/1.1".format(host, date, method, path)
    signature_sha = hmac.new(api_secret.encode('utf-8'), signature_origin.encode('utf-8'),
                             digestmod=hashlib.sha256).digest()
    signature_sha = base64.b64encode(signature_sha).decode(encoding='utf-8')
    authorization_origin = "api_key=\"%s\", algorithm=\"%s\", headers=\"%s\", signature=\"%s\"" % (
        api_key, "hmac-sha256", "host date request-line", signature_sha)
    authorization = base64.b64encode(authorization_origin.encode('utf-8')).decode(encoding='utf-8')
    values = {"host": host, "date": date, "authorization": authorization}
```

```
                return requset_url + "?" + urlencode(values)


# 配置 API 请求的 URL 和请求体内容
url = 'https://itrans.xf-yun.com/v1/its'
body = {
"header": {"app_id": appid, "status": 3, "res_id": RES_ID},
"parameter": {"its": {"from": "en", "to": "cn", "result": {}}},
"payload": {"input_data": {"encoding": "utf8", "status": 3, "text": base64.b64encode(TEXT.encode("utf-8")).decode('utf-8')}}
}

# 调用 assemble_ws_auth_url 函数,生成完整的带鉴权的 API 请求 URL
request_url = assemble_ws_auth_url(url, "POST", api_key, api_secret)

# 发送 HTTP POST 请求
headers = {'content-type': "application/json", 'host': 'itrans.xf-yun.com', 'app_id': appid}
response = requests.post(request_url, data=json.dumps(body), headers=headers)

# 解析并输出翻译结果
tempResult = json.loads(response.content.decode())
print(base64.b64decode(tempResult['payload']['result']['text']).decode())
```

运行代码,生成内容如下:

```
{"from":"en","to":"cn","trans_result":{"src":"Machine translation has made great progress in recent years, but there is still a lot of room for improvement.","dst":"近年来,机器翻译取得了长足的进步,但仍有很大的提升空间。"}}
```

代码中"en""cn"分别代表"英语"和"汉语",若需改变翻译语言,修改代码中"text"变量的内容,同时改变代码中"from"和"to"语言设置,具体语言代码参见讯飞开放平台相关页面。

星火大模型在与中文相关的翻译任务中表现出色,对中文的理解和生成能力很强,能够准确把握中文的语言特点和文化内涵。作为一个大型语言模型,它具有强大的上下文理解能力,能够提供更加连贯和符合语境的翻译。它集成了讯飞在语音识别和合成方面的优势,可以提供更全面的多模态翻译服务,如语音到文本的翻译。星火大模型是通用的大型语言模型,而非专门为翻译任务优化的系统,在某些高度专业化的领域可能不如专用的翻译引擎或翻译工具准确。相比于一些国际知名的翻译系统,星火模型在小语种或非常专业领域的翻译质量还有改进空间。和其他大型语言模型一样,星火模型在翻译过程中可能会产生"幻觉"。作为一个相对较新的系统,它在某些语言对或特定领域的翻译性能还需要更多的实际应用检验。

参 考 文 献

[1] BIRD S, KLEIN E, LOPER E. Natural Language Processing with Python[M]. Sebastopol: O'Reilly Media, Inc. 2009.
[2] Python 官网. https://www.python.org/.
[3] PERKINS J. Python 3 Text Processing with NLTK 3 Cookbook[M]. 2nd ed. Birmingham: Packt Publishing Ltd. 2014.
[4] PERKINS J. Python Text Processing with NLTK 2.0 Cookbook[M]. Birmingham: Packt Publishing Ltd. 2010.
[5] SARKAR D. Text Analytics with Python: A Practical Real-world Approach to Gaining Actionable Insights from Your Data[M]. New York: Apress. 2016.
[6] WES MCKINNNEY W. Python for Data Analysis: Data Wrangling with Pandas, Numpy, and Ipython[M]. 2nd ed. Sebastopol: O'Reilly Media, Inc. 2017.
[7] ZHANG P, PAN Y. A comparative study of keywords and sentiments of abstracts by python programs[J]. Open journal of modern linguistics, 2020(10): 722-739.
[8] 管新潮. Python 语言数据分析[M]. 上海：上海交通大学出版社，2021.
[9] 胡显耀. 语料库文体统计学方法与应用[M]. 北京：外语教学与研究出版社，2021.
[10] 刘华. 语料库语言学：理论、工具与案例[M]. 北京：外语教学与研究出版社，2020.
[11] 柳毅. Python 数据分析与实践[M]. 北京：清华大学出版社，2019.
[12] 魏伟一，李晓红. Python 数据分析与可视化[M]. 北京：清华大学出版社，2020.